U.S. Department of Energy
Energy Efficiency and Renewable Energy
Bringing you a prosperous future where energy is clean, abundant, reliable, and affordable

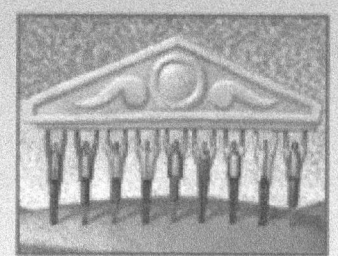

Commissioning for Federal Facilities

A practical guide to building commissioning, recommissioning, retrocommissioning, and continuous commissioning

Developed by the U.S. Department of Energy,
Federal Energy Management Program, and
Enviro-Management & Research, Inc.

The Right Answers. Right Now.

Chapter 1
Introduction

Building commissioning has often been likened to commissioning of a ship, where the Owners thoroughly verify and prove the functional performance of all parts – engines, compasses, sonar, radar, radio, generators, potable water systems, and so on – under all possible conditions and as a condition of acceptance before placing the ship in service. And where the Owner checks the presence of system operating and procedures manuals and the availability of up-to-date navigation charts. And where the crew has been properly and thoroughly trained on the ship's systems' operations and emergency procedures. Commissioning is not new – ships and aircraft have been commissioned for years.

Building commissioning has its roots in the Quality Control programs of the 1970s and is a direct product of the Total Quality Management programs of the 1980s. Commissioning is a direct response to building Owners who complain that their facilities do not meet performance expectations, are extraordinarily expensive to operate and maintain, lack valuable documentation, and are staffed by personnel who are unfamiliar with and have never been trained on the building's highly complex operations and control systems.

Until now, many of us thought of building construction completion and turnover as physically completing an installation, throwing the switch, making a few adjustments, spending minimal time with the operators by pointing to the equipment with one hand and a manufacturer-supplied operations manual (that may or may not match the specific equipment) with the other, then

"Building commissioning has often been likened to the commissioning of a ship."

1

Goals of Commissioning:

◆ Provide a safe and healthy facility.

◆ Improve energy performance and minimize energy consumption.

◆ Reduce operating costs.

◆ Ensure adequate O&M staff orientation and training.

◆ Improve systems documentation.

walking away. We would return only when the operating personnel or owner complained.

A successful project was not necessarily the one with the most satisfied client, optimal indoor environment, most reliable and efficient operation, or that would have had the lowest possible operating and maintenance costs. Typically, it was the one with the fewest extras and change orders and the one with the shortest punch list.

Further, the construction budget and operating budget came from differ-

ent sources and programs. So once the construction was completed and beneficially accepted, the building was handed off as rapidly as possible, leaving building maintainers to struggle with any residual construction or operational problems. In all fairness to the constructors, by this time the Owner was usually pressing to move into the building, either oblivious to or willing to accept the risks associated with a potentially problematic facility.

Most existing buildings have never undergone a formal commissioning or quality assurance process. Many buildings are limping along inefficiently in terms of performance. Owners are unaware of deficiencies as long as the building is reasonably comfortable and occupant complaints do not reach a crescendo.

In reality, the building systems may be becoming increasingly unreliable and inefficient through design, ineffective maintenance and operations procedures, outdated technologies, insufficient training, occupant habits, mission changes, environmental changes, workplace configurations, and more.

All of that has now changed with commissioning. As described by the Canadian Department of Public Works, buildings now "leave port" only when they are fully operational, function as the owner intended, are fine-tuned for maximum performance, staffed with "crews" who are fully trained in the regular and emergency operation of the facility, and furnished with a complete set of relevant operations, maintenance,

Guidebook Objectives

◆ Provide an introduction to commissioning approaches to a variety of professionals involved with the management, operation, and maintenance of Federal buildings.

◆ Illustrate case histories, including cautionary lessons learned.

◆ Provide guidance on commissioning best practices.

◆ Demonstrate how commissioning can help Federal facility managers meet energy efficiency goals and LEED certification requirements.

◆ Demonstrate how commissioning can be integrated in facility management and O&M programs to make those programs more efficient and effective.

◆ Demonstrate how different types of commissioning (such as retrocommissioning and continuous commissioning) can be incorporated into a variety of building types and applications, above and beyond the most commonly understood commissioning approaches.

facility intent and design, and emergency procedures documentation.

So... What Is It?

Commissioning is a method of risk reduction.

The National Conference on Building Commissioning has established an official definition of total building commissioning as follows:

> "Systematic process of assuring by verification and documentation, from the design phase to a minimum of one year after construction, that all facilities perform interactively in accordance with the design documentation and intent, and in accordance with the owner's operational needs, including preparation of operational personnel."

Total or *whole* building commissioning differs from "building commissioning" inasmuch as the former refers to the whole process from the project planning to post-acceptance, as well as to all of the building systems that are integrated and impact on one another, such as HVAC, lighting, electrical, plumbing, building envelope and their respective controls and technologies.

Building commissioning that is not qualified as total or whole building commissioning may be more selective in terms of the phases during which the commissioning activities actually take place (e.g., the Commissioning Agent may be hired to commence work late in the design or during the construction phase) or in terms of the systems to be commissioned (e.g., HVAC and electrical systems only). It is essentially a subset, or a slice of the whole building commissioning pie, and for the purposes of this document, the terms will be used interchangeably.

What Are The Goals?

The goals of commissioning are to:

- Provide a safe and healthy facility.
- Improve energy performance and minimize energy consumption.
- Reduce operating costs.
- Ensure adequate O&M staff orientation and training.
- Improve systems documentation.

It's purpose, however, is to provide a framework for a quality-oriented **team effort** that **reduces project costs** while delivering system **reliability** and **quality**. Thereby, it enhances long-term value to the Owner.

Why Do We Do It?

Following the David Letterman model, the following are the top ten reasons why people commission:

10. For the documentation

9. To ensure integration of building systems

8. To prevent premature failure

7. For the transfer of knowledge

to building operators and engineers

6. For the performance testing of complex systems

5. To ensure equipment accessibility

4. To improve energy performance

3. For improved system and equipment reliability

2. For project cost control

1. To meet Owner expectations

Owners use commissioning's systematic, documented, and collaborative process to ensure that a building and its components' systems will:

■ Have high quality, reliability, functionality, and maintainability;
■ Meet energy and operational efficiency goals;
■ Operate and function as the owner intended and as designed; and
■ Be what the Owner paid for.

These objectives are achieved by verifying that the equipment performance meets or exceeds the designer's intent as documented in the project drawings, specifications, and design intent documentation.

From the aspect of energy savings, commissioning has proven itself time and again. In existing buildings, whole-building energy savings average about 15 percent at a cost of about $0.27 per square foot and with a payback of about 8.5 months. In new construction, commissioning costs about $1.00 per square foot and pays back within about 4.8 years.

In addition, consider the cost savings associated with worker productivity, detection of failed parts and impending failure, and other benefits not included in these savings. These

---

A major university commissioned six major buildings totaling 260,000 square feet. More than 500 "completed" variable air volume (VAV) boxes were tested with the following results:

◆ Nine were installed without the main supply air connected

◆ 52 had control programming problems

◆ 23 had control valve problems (including above-ceiling actuators not connected)

◆ 25 could not achieve the maximum air flow recorded by the balancer (e.g., frozen dampers)

◆ Eight thermostats were in poor locations, such as near diffusers and heat generating sources

(Source: S. Angle, *Engineered Systems,* January 2000.)

numbers will be addressed again later in greater detail.

Examples of common problems that commissioning addresses that drive energy costs up but may or may not cause discomfort or other visible problems include:

- Outside air dampers stuck in the always open or always closed position.
- Adjustable speed drives that no longer adjust properly.
- Unconnected flexible ductwork.
- Malfunctioning control systems components that do not properly respond to their prescribed control sequences.
- Incorrect sequences of operation.
- Energy management systems that have not been updated to reflect system modifications.
- Changed facility uses that affect personnel loading and partition configuration changes that affect air flow.
- Controls sensors that are out of calibration.
- Controls that are permanently overridden.
- Heating and cooling systems that fight each other.
- Thermostats and other control devices that are improperly placed.

How Do We Do It?

Quality control has historically been associated with static and individual systems, such as piping, ductwork, building aesthetics, air handlers, and other standard punch list items. The project inspector ensures material

and workmanship quality, technical specifications adherence, and code compliance. Quality control ensures the installation will pass specified tests (such as start-up, operating, hydraulic, and leakage tests), and ultimately, pass the final punch list.

Commissioning is usually associated with dynamic and integrated mechanical, electrical, security, life-safety, conveyance, and other systems and their controls. Today's use of commissioning recognizes the integrated nature of all building systems' performance. Top concerns are security, indoor air quality, and integrated life-safety. It also takes a proactive approach toward the operation and maintenance of the installed system.

In addition to ensuring that a system is delivering the required flow and pressure, commissioning tests the entire integrated system from controls to delivery; tests the interoper-

Commissioning objectives are met by verifying that the equipment performance meets or exceeds the designer's intent as documented in the project drawings, specifications, and design intent documentation.

5

ability between systems; tests the condition and operation of key components; ensures the completeness and quality of O&M manuals and skills training; is mindful of maintainability, accessibility, supportability, and reliability issues; and documents the entire process. Typically, these are not high priority issues in a standard quality control program. However, by design, commissioning includes these so that there is a high degree of confidence that the building's systems have been installed correctly and will operate as required.

WHEN DO WE DO IT?

The widely held misconception is that commissioning is checking off the installation and start-up menu provided by the equipment manufacturer. In reality, commissioning is results-oriented, comprehensive, and emphasizes communication, inspection, testing, and documentation. When properly executed, commissioning begins with pre-design planning, continues into post-occupancy,

and is heavily involved in the planning, design, construction, and acceptance stages in between.

In existing buildings that have never been commissioned before, **retro**commissioning can take place at anytime, unless the facility and/or major equipment are programmed for replacement in the immediate future. In that case, it is usually advantageous to wait and commission the facility as part of the construction effort. Otherwise, commissioning an existing building will likely uncover a multitude of deficiencies that affect the building's efficiency and ability to operate as required.

In existing buildings that have been previously commissioned, **re**commissioning is usually recommended at about the 3-5 year point since the previous commissioning. However, the most proactive programs commission their buildings continuously, using and trending data from their building management systems, installed meters and sensors, and even utility data. In these cases, commissioning never really stops, as analysis is conducted continually to detect impending failures, abnormalities, and efficiency opportunities.

WHO DOES IT?

The Federal Government is in the forefront of commissioning. The Government's landlord, the General Services Administration, now requires *all* GSA capital improvement projects to employ Total Commissioning practices as addressed in its *Building Commissioning Guide*.

6

The requirement is in GSA's design criteria document, *Facilities Standards for the Public Buildings Service* (P-100). All new construction for GSA must now employ commissioning, beginning with the project planning phase and concluding with the post occupancy evaluation phase. The cost for commissioning is included as a line item in the construction project budget. Other Federal Agency real property owners will be establishing similar requirements to at least some extent, if they have not already.

Grants and special incentives are available for Owners considering commissioning their facilities. These are typically available from Federal and state entities, such as the New York State Energy Research and Development Authority (NYSERDA). For example, NYSERDA and the Department of Energy provide a no cost, risk-free scoping study to Federal building operators to determine the cost effectiveness of commissioning specific existing buildings. Some utility companies also provide rebates to Owners conducting commissioning in new and existing buildings, particularly if LEED certification is achieved.

STUDY QUESTIONS

1. How does commissioning change the traditional definition of a "successful" construction project?

2. What are the top ten reasons why people employ commissioning, and how could your facility benefit from these reasons?

3. What is the difference between commissioning, and *total* or *whole* building commissioning?

4. How is the commissioning process different from a quality control process?

5. GSA is requiring commissioning to be implemented on its new construction projects. Is commissioning required by your Agency or organization? How important is commissioning in your Agency or organization?

8

This page left blank intentionally

Chapter 2
Types of Commissioning

The first step in considering or planning a commissioning program for your facility is to understand the different types of commissioning available, and which types of commissioning are best suited to your facility's unique requirements. In general, a commissioning program is best applied during the following:

- During new construction or a major renovation of an existing building.

- When an existing building is experiencing problems with operational performance, energy efficiency, and/or occupant comfort and safety.

- As a maintenance approach to ensure that equipment and systems are operating at peak performance, energy efficiency is optimized, and occupant comfort and safety are high.

The types of commissioning that fit into these applications that will be discussed in this and subsequent chapters are:

- Commissioning for New Construction/Renovation

- Retrocommissioning
- Recommissioning
- Continuous Commissioning

COMMISSIONING FOR NEW CONSTRUCTION / MAJOR RENOVATION

Commissioning is a systematic process of ensuring that all building systems perform interactively according to the design intent and the Owner's operational needs. The process evaluates building equipment, subsystems, operation and maintenance (O&M) procedures, and performance of all building components to ensure that they function efficiently, and as designed, as a system. This is achieved by beginning in the planning or early design phase of a construction project with the documentation of design intent, and continuing through construction, acceptance, and

"A leader takes people where they want to go. A great leader takes people where they don't necessarily want to go but ought to be."

Rosalynn Carter

9

In this Chapter

◆ Commissioning for New Construction / Major Renovation
◆ Retrocommissioning
◆ Recommissioning
◆ Continuous Commissioning
◆ Best Practices

the warranty period with the actual verification of each building system's performance.

The commissioning process encompasses and coordinates the traditionally separate functions of system documentation, equipment startup, control system calibration, testing and balancing, performance testing, and training. It defines a maintenance baseline against which future condition assessments and trending can be compared.

Commissioning may include the building envelope, the building HVAC systems, controls, electrical, conveyance, plumbing fixtures, life safety, security, or any combinations of these systems and others.

The specific person or organization that conducts and oversees the commissioning process is the Commissioning Authority (or Agent), commonly referred to as the "CxA."

Often, the assistance of subject matter experts is required. Commissioning of laboratories requires special attention and involvement of the Owner's environmental health and safety (EH&S) staff as part of the commissioning team. For example, they will help the CxA understand

the containment goals for fume hoods and bio-safety cabinets and of primary and secondary barriers so that their compliance with the requirements can be verified during the commissioning process.

Often too, local fire marshals alone are responsible for the inspection, testing, and approval of all fire prevention and protection devices and systems. In that case, commissioning is coordinated with the fire marshal's work, his efforts are observed by the CxA, and a copy of the official fire marshal report is included as part of the Final Commissioning Report.

The commissioning process does not take away from or reduce the responsibility of the system designers or installing contractors to provide a finished and fully functioning building. Commissioning does not take the place of or reduce in any way the contractor's responsibilities for conducting an active project quality control program.

The Commissioning Process

Commissioning is systematic. It includes testing all items in all modes of operation. Equipment is first inspected while it is turned off to make sure that it is installed fully and correctly. Equipment is then energized, started, and tested under controlled conditions. After this initial testing and inspection, integrated systems are tested through all cycles and scenarios, including power failure and emergency alarm modes, to ensure they operate as required and intended.

In the course of commissioning, key parameters and baseline information of the systems are documented, organized, and preserved in the Commissioning Report and O&M manual, as applicable.

Commissioning typically follows the phases of the new construction or renovation project. Although it is not necessary to perform commissioning tasks during each phase of construction, implementing the process throughout the life of the project will produce the best results. Each of these will be discussed in greater detail later:

Pre-design

- Determine project objectives and develop Owner's Criteria.
- Develop commissioning requirements.
- Hire or assign Commissioning Authority (CxA).

Design

- Design team develops project design; CxA reviews design intent, basis of design documents, and drawings and provides feedback to design team.
- CxA develops commissioning plan.
- Design team develops project specifications; CxA develops

What Type of Commissioning Should I Choose?

My building is...	Consider...
... going to be undergoing a major renovation in the next year.	**Commissioning** - Ideal for new construction or major renovation, and best implemented through all phases of the construction project.
... old and experiencing a lot of equipment failures.	**Retrocommissioning** - Ideal for older facilities that have never been through a commissioning process.
... relatively new and was commissioned during its construction, but our energy costs have been climbing recently.	**Recommissioning** - Ideal to tune up buildings that have already been commissioned, bring them back to their original design intent and operating/energy efficiency
... large and complex. We have a metering system and a preventive maintenance program, but will still struggle with high energy costs and tenant complaints.	**Continuous Commissioning** - Ideal for facilities with building automation systems (BAS), advanced metering systems, and advanced O&M organizations.

Commissioning's Objectives

Commissioning is intended to achieve the following specific objectives:

Verification...

◆ ... that applicable equipment and systems are installed according to the manufacturer's recommendations and to industry accepted minimum standards

◆ ... that applicable equipment and systems receive adequate operational checkout by installing contractors

◆ ... and documentation of proper performance of equipment and systems under various conditions

◆ ... of the proper interactions between systems and subsystems

◆ ... that systems and O&M documentation left on site is complete

◆ ... that the building's O&M staff has been adequately trained.

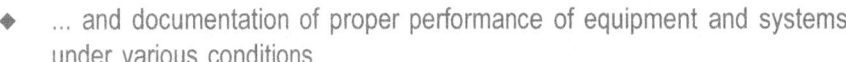

commissioning specifications.

Installation/Construction

■ CxA gathers and reviews design and project documentation.

■ CxA holds periodic commissioning meetings to integrate the process and schedule into the overall construction project.

■ CxA develops verification checklists and functional performance test (FPT) forms.

■ CxA monitors construction progress.

■ CxA works with the Owner to ensure selected maintenance staff members are given the training opportunity of observing the installation and testing of specific systems for which they will inherit maintenance responsibilities.

■ CxA works with installing contractors to verify start-up and perform verification to ready systems and equipment for FPT.

Acceptance

■ CxA directs and oversees installing contractors' performance of FPT, observed by Owner's selected maintenance staff; deficiencies are reported.

■ CxA validates building Testing and Balancing (TAB) report data.

■ CxA directs and oversees installing contractors' performance of equipment condition acceptance testing, observed by Owner's selected maintenance staff; deficiencies are corrected and condition baseline data is included in the final Commissioning Report and O&M manual.

- CxA works with installing contractors to reschedule FPT as needed when deficiencies are present; corrections to deficiencies are verified by CxA.
- CxA reviews and verifies O&M manuals for completeness and applicability.
- CxA oversees, Contractor conducts, and Owner coordinates prescribed training for the O&M staff.

Post-acceptance/Warranty

- CxA prepares the Final Commissioning Report.
- CxA conducts site visits to interview O&M staff on system performance.
- Deferred and/or seasonal testing is performed.

Types of Testing Used

Verification Checks

Verifications checks are equipment inspections that ensure proper installation and configuration. This testing employs checklists to verify that the equipment or system is ready for initial start-up (e.g., flexible conduit is connected, belt tension is correct, oil levels are adequate, labels are affixed, gauges are in place, and sensors are calibrated). Some verification checklist items entail the simple testing of the function of a component, a piece of equipment, or system (such as measuring the voltage imbalance on a three-phase pump motor of a chiller system).

For most equipment, the installing contractors execute the checklists on their own. The CxA requires that the procedures and results be documented in writing and usually witnesses only the verification testing of the larger or more critical pieces of equipment. Other components are validated randomly by the CxA.

Functional Performance Tests

Functional performance tests are a series of tests of the function and operation (and sometimes, condition) of equipment and systems using manual (direct observation) or monitoring methods. Functional performance testing is the dynamic testing of systems (rather than just components) under full operation (e.g., the chiller pump is tested interactively with the chiller functions to see if the pump ramps up and down to maintain the differential pressure set point).

Systems are tested under various modes, such as during low cooling or heating loads, high loads, component failures, unoccupied condition, varying outside air temperatures, fire alarm, and power failure. The systems are run through all the control system's sequences of operation, and components are verified to respond as the prescribed sequences state. The CxA develops the functional test procedures in a sequential written form, coordinates, oversees, and documents the actual testing, which is usually performed by the installing contractor or vendor.

In addition, seasonal functional performance tests may also be performed, during which the installing contractor and/or CxA performs the functional performance test during different seasonal weather conditions.

Functional performance testing also may include procedures for condition acceptance testing. Condition acceptance testing uses condition monitoring techniques, usually associated with reliability centered maintenance, to identify latent manufacturing, transportation, and installation defects that affect the condition of the equipment at the time of acceptance.

The most common techniques will use vibration analysis to inspect for mechanical alignment and balance, softfoot, and internal and bearing defects; infrared thermography to determine the presence of high resistance and other problematic electrical connections; ultrasound to determine the presence of fluid (e.g., compressed air, steam, gas) leaks; lube oil analysis to determine the quality, condition, and appropriateness of lubricating oils and their additives; and/or motor testing and electrical testing, where the condition of the insulation is of major importance.

Not all commissioning programs include condition acceptance testing. However, there is no better time to determine the *physical hidden condition* of the equipment (while functional performance testing looks at *operating parameters*) than as a condition for acceptance while the warranties are still active and to establish the condition baseline for the ensuing maintenance program.

Testing and Balancing (TAB)

System testing and balancing may or may not be included as part of the commissioning (that is, the TAB technicians may or may not work for the CxA). However, validation of the TAB results by random spot checking actual output against the documented TAB data normally will be included in the commissioning process regardless of the TAB contractor's relationship within the commissioning team.

Advantages

■ Commissioning leads to improved system performance by ensuring that equipment and systems are properly designed, installed, maintained, and optimized to work together.

■ Commissioning can reduce change orders and improve contractor performance and awareness. Testing and monitoring make contractors more aware of the quality of their work.

■ Commissioning can improve the overall construction process and project turnover. The process provides for better project communication and enhanced conflict resolution. Commissioning also provides for follow-up site visits to address any problems that may occur after project turnover.

Functional performance testing determines the operating parameters of equipment and systems, while condition acceptance testing determines the physical hidden condition.

14

Double Checking the TAB Report

In a newly constructed health sciences laboratory and classroom facility at a major university, the CxA performed a random validation check of the testing and balancing contractor's TAB report. Starting with a random check of 10% of the air registers, the CxA found an inordinate number of differences between the actual and TAB-recorded readings. The CxA increased the sample to 25% and found an even greater difference. Further investigation found that the TAB contractor failed to accurately test and balance the air and water system at all and fraudulently recorded made up numbers on the official TAB report.

The contractor paid heavily as a result. The TAB was re-performed correctly by a reputable contractor. The project acceptance was delayed for several weeks as a result of the required re-work. However, because of the CxA's testing and verification, the Owner ended up with a fully and properly functioning and balanced HVAC system that would probably not have been realized until well after the facility became occupied, occupant complaints drove a costly investigation, and payment had already been made for the original, fraudulent TAB work.

- A reduction in TAB related to construction/major renovation costs can occur because systems and equipment are more likely to be properly balanced during start-up and verification checks.

- Studies show that commissioned buildings typically save 10 to 20 percent of utility costs compared to similar non-commissioned buildings by working to ensure that system components operate together most efficiently.

- Commissioning saves energy and environmental emissions. It is a required factor for points toward Leadership in Energy and Environmental Design (LEED) certification.

- Commissioning ensures that a building is pressurized and has correct fresh air changes for indoor air quality (IAQ). This decreases mold-related problems and "sick building" syndrome. Improved IAQ also impacts the Owner's liability relative to occupant health and comfort and increases worker productivity.

- Commissioning has been shown by the insurance industry to reduce liability and losses related to fire and wind damage, ice and water damage, power failures, professional liability, and health and safety. Reduced risk and liability can also increase the asset value of the building.

- It is much easier and less expensive to maintain a building that operates correctly than to maintain one that does not. Designs that have been reviewed for maintainability and sustainability, and equipment that has been installed and tested properly and optimized for maximum efficiency, will experience fewer problems and require less unscheduled O&M time.

■ Equipment condition-accepted during commissioning verifies the equipment condition prior to expiration of its warranties and provides a condition baseline for the ensuing maintenance program.

■ Commissioning can extend equipment life and reduce warranty claims, leading to fewer warranty claims, service calls, reduced energy use, and reduced potential for catastrophic equipment failure.

■ Commissioning provides more useful O&M condition baseline and performance data that is specific to the systems and equipment installed. It details the way the equipment should be operated, outlines preventive maintenance procedures and schedules, and provides information on warranties, vendor points of contact, and spare parts.

■ The maintenance staff is trained on site by observing the work as it progresses as well as by formal instruction customized to the specific equipment and systems installed.

■ Commissioning addresses common occupant concerns such as thermal comfort, air flow and air quality, and lighting levels to ensure that occupants are comfortable, safe, and productive in their work spaces.

Disadvantages

■ The first costs of commissioning are construed by Owners as being high only to ensure that the contractor's work is of a quality that he's already contracted to perform. There is little quantifiable data on the potential cost savings (both energy and operational) that the commissioning process will generate for the specific, as-yet operational building. Nor is there any way to benchmark in advance, energy and operational performance in the case of new construction (in which the "existing" conditions do not yet exist).

■ There is no guarantee of savings. The commissioning process is designed to optimize all building system and equipment operations to meet the design intent; most of the savings occur through avoided costs.

■ If a quality assurance program is already utilized by the A/E, construction manager, and installing contractors, commissioning may be perceived to be redundant and/or confrontational.

RETROCOMMISSIONING

Retrocommissioning is a systematic process for improving and opti-

16

mizing building performance. Retrocommissioning applies to exiting buildings that have never gone through any type of commissioning or quality assurance process. Its focus is usually on energy-using equipment such as mechanical equipment, lighting, and related controls.

Like commissioning, retrocommissioning is concerned with how equipment, systems, and subsystems function together, but it does not generally take a whole-building approach to efficiency. The process can identify and solve problems that occurred at construction, but also addresses problems that have developed to this stage in the building's life. And while the goal of retrocommissioning may be used to bring the building, its systems, and equipment back to its original design intent, this is not a requirement. The original design intent documentation may be lost or no longer relevant.

The Retrocommissioning Process

Retrocommissioning is not tied to a specific new construction or renovation project, and therefore does not necessarily follow the same process as commissioning.

Retrocommissioning typically follows a four-part process:

1. Planning
   - Identify project objectives.
   - Decide which building

17

Type of Commissioning	Why?	Who?	When?	How?
Commissioning	Ensure that the building and its systems and equipment operate as designed	Independent CxA hired by the Owner or the project Construction Manager	Once, during new construction or renovation	Verification and functional performance testing
Retrocommissioning	Identify and correct problems and optimize performance	Facility O&M staff or independent CxA	Once, in response to specific problems or to establish a commissioning program	Diagnostic monitoring and functional performance testing
Recommissioning	Ensure that the building and its systems and equipment continue to operate as designed, or meet current operating needs	Facility O&M staff or independent CxA	Periodically as the building ages, or ongoing as part of the facility O&M program	Functional performance testing
Continuous Commissioning	Identify and correct problems and optimize performance	Facility O&M staff or independent CxA	Ongoing as part of the facility O&M program	Data monitoring and trending

systems should be analyzed for improvements.
- Define tasks and assign responsibilities.

2. Investigation
- Determine how the selected systems are supposed to operate, or how they could operate more efficiently.
- Perform tests to measure and monitor how the targeted systems currently operate.
- Prepare a prioritized list of the operating deficiencies found and recommended corrective actions.

3. Implementation
- Correct operating deficiencies (highest priority to lowest).
- Perform tests to verify proper and/or improved operation.

4. Hand-off
- Prepare a report of improvements made.
- Provide training and documentation on how to sustain proper and/or improved operation.

Types of Testing Used

The investigation phase of retrocommissioning involves review of current O&M practices and service contracts, spot testing of equipment and controls, and trending or electronic data logging of pressure temperatures, power, flows, and lighting levels and use.

In addition, both diagnostic monitoring and functional performance tests are performed to determine temperatures, critical flows, pressures, speeds, and currents of the system components under typical operating conditions. Typical diagnostic monitoring methods employed include energy management control system (EMCS) trend logging and stand-alone portable data logging. The retrocommissioning process involves development of a diagnostic monitoring plan and functional performance test plan, both including test forms.

Advantages

The advantages of retrocommisioning are nearly the same as those of commissioning:

- Improved system performance

- Energy savings and optimal energy efficiency (commissioning is a required factor for points toward LEED-EB certification).

- Improved indoor air quality and reduced liability.

18

- Increased occupant comfort, safety, and productivity.

- Reduced O&M costs and fewer service calls.

- Extended equipment life and reduced warranty claims.

- Increased system reliability.

- Increased O&M training and improved documentation.

In addition, a retrocommissioning program can result in increased interest in facilities improvement and asset management at all levels. Commissioning can also complement an ongoing facilities management process improvement program.

Disadvantages

Retrocommissioning and commissioning also share many of the same disadvantages:

- The first costs of retrocommissioning may be considered by the Owner to be high and unaffordable since it usually must compete with other priorities from the facility's operating budget. To counter this perception, retrocommissioning should be "sold" to Management as a profit center by demonstrating estimated energy, maintenance, and productivity savings that will result.

- Savings may not be the primary focus. The retrocommissioning

process is designed to optimize building system and equipment operations to meet the design intent or current building requirements. There is no *guarantee* of savings, however they are still a likely by-product that occurs through avoided costs.

- There is a significant up-front workload when performing retrocommissioning for the first time. Documentation, including diagnostic test forms and functional performance test forms, that does not exist on site must be compiled and/or developed.

RECOMMISSIONING

Recommissioning refers to commissioning of an existing building that has already gone through the commissioning process. Why the need to commission again, particularly if the building was commissioned during its construction or a recent major renovation? Recommissioning provides additional opportunities to improve facility efficiency and addresses issues that may have arisen since the original commissioning, such as:

- Changes in the purpose or occupancy of the facility that have occurred since the building was first commissioned.

- Changed building configurations and occupancy patterns since the building was first commissioned (e.g., is an original laboratory now being used for

Based on three years of data, a demonstrable 10-percent reduction in energy use qualifies for 1 LEED-EB point; a 20-percent reduction qualifies for 2 points; and a 30-percent reduction qualified for 3 points (maximum)

19

When was the last major equipment change-out? Is existing equipment relatively new, or at the end of its useful life? Is a major renovation on the horizon? These questions will help you decide whether or not your facility can benefit from existing building commissioning.

storage or conference room space?).

■ New higher efficiency systems and equipment that have become available since the building was first commissioned.

■ Persistent high energy costs despite efforts made to monitor and control energy consumption and demand.

■ Equipment and systems that do not operate optimally, or too often fail, despite a strong facility operations and maintenance program.

■ Technologies are now available that improve energy and operational efficiency, but there is little or no money programmed in the capital improvement budget.

■ Federal statute, Executive Order, or other requirement mandates that efforts be taken to achieve better energy and water savings and healthier indoor environments.

■ National recognition for energy and emissions reduction is being sought through the LEED-EB Green Buildings program.

Like commissioning, recommissioning involves functional performance testing of most or all major building systems including HVAC, building automation, lighting, life safety, and conveyance. Mechanical systems operation and controls are most closely scrutinized because they of-

ten are the source of the biggest operational problems and are thus likely to produce the biggest cost savings. Results of testing are documented, and recommendations for improving performance are implemented.

The Recommissioning Process

During recommissioning, the tests that were performed during the original commissioning are performed again, with the goal of ensuring that the building is operating as designed or according to newer operating requirements. The development of new project documentation and testing procedures and forms is not required. However, these documents can be updated if the building and its systems and equipment have changed dramatically since the original commissioning.

Recommissioning can be undertaken as an independent process in response to a specific requirement or concern (such as those listed above), or periodically scheduled as part of the building's operations and maintenance program. In general, the more substantial changes that a facility goes through, the more often it should be recommissioned if a continuous commissioning program is not in place. If there are no known substantial changes to the facility and its operation, it is recommended in general that the facility be recommissioned every 3-5 years.

An independent CxA can be hired to perform recommissioning, or the facility O&M staff can use the existing test forms to perform recommissioning in house.

Types of Testing Used

Recommissioning starts with:

- Site observation;
- Interviews with occupants;
- Analyses of energy metering data (if available);
- Review of current O&M practices and service contracts;
- Spot testing of equipment and controls; and
- Trending or electronic data logging of pressure, temperatures, power, flows, and lighting levels and use to determine current conditions (this replaces verification checks).

Recommissioning then uses the same functional performance test forms that were developed during the initial commissioning process to test systems dynamically under full operation. Systems are tested under various modes, such as during low cooling or heating loads, high loads, component failures, unoccupied conditions, varying outside air temperatures, fire alarm, and power failure. The systems are run through all the control system's sequences of operation. Components are checked for their responsiveness to the prescribed sequences and validated.

Unlike commissioning, the bulk of the functional performance testing performed during recommissioning may be carried out by the building O&M staff.

Advantages

In addition to the advantages listed under commissioning and retro-commissioning, recommissioning also provides the following:

- Periodic recommissioning can contribute to the persistence of commissioning savings and benefits, and will ensure that the building and its equipment and systems remain in compliance with original design intent.

- Recommissioning affords facility managers the opportunity to update building, system, and O&M documentation and to modify the design intent, if necessary, to reflect changes in building requirements.

- Functional performance test forms have already been developed and are ready for use.

- Recommissioning can increase O&M knowledge and skills in diagnosing operating problems and determining and implementing corrective strategies.

- Recommissioning can identify problems not readily apparent due to long-term storage of equipment, such as breakdown of dielectrics, degraded fluids, failed batteries, leaking seals, and flattened bearings.

Disadvantages

- Recommissioning may be an

21

occasional event and may take place many years after the initial commissioning, depending on the level of Management support and the availability of funds.

- Recommissioning is often implemented only in response to breakdowns in equipment or systems, and not as a proactive tool to ensure building performance optimization.

- Recommissioning should not be used as a substitute for major equipment change-out or systems redesign that may, in fact, be required.

- There is a risk of facility Management adhering to an outdated design intent rather than updating the design intent for a building's current requirements.

CONTINUOUS COMMISSIONING

Continuous commissioning, like retrocommissioning, is a systematic approach to identifying and correcting building system problems and optimizing system performance in existing buildings. Any similarities between the programs end there, however. Continuous commissioning is distinct because its primary focus is on ensuring the persistence of building systems optimization. It is an ongoing process for existing buildings employed to resolve operating problems, improve building comfort and safety, optimize energy use, and identify retrofits.

Continuous commissioning requires benchmarking of energy use using available installed building automation systems and metering equipment. Data are continuously gathered and compared against the benchmark to measure building efficiency and ensure that equipment and systems operate at optimal levels throughout their useful lives.

While continuous commissioning is closely related to (and often integrated into) a facility operation and maintenance program, it goes beyond O&M to ensure that the building and systems operate optimally to meet current requirements, evaluating both building functionality and equipment and system functions.

Continuous commissioning can be provided by a qualified third party CxA, or by well-trained members of the O&M staff.

The Continuous Commissioning Process

Continuous commissioning is accomplished in two phases: project development, and implementation and performance verification. During project development, the CxA or O&M team screens potential project targets, performs a project audit, and develops the project scope.

During the second phase, the CxA or O&M team:

- Develops the commissioning plan and forms the project team
- Develops performance baselines
- Conducts system measurements

Retrocommissioning, Recommissioning, Continuous Commissioning... I'm Confused!

All three terms apply to commissioning of existing facilities, and all three aim to improve operating performance, energy efficiency, and occupant comfort and safety. Here's how they're different:

Retrocommissioning

- A one-time event
- The building *has not* been previously commissioned
- May or may not adhere to building's original design intent
- Utilizes diagnostic monitoring and functional performance tests

Recommissioning

- A one-time, periodic, or occasional event
- The building *has* been previously commissioned
- Adheres to building's original design intent
- Utilizes previously developed functional performance tests

Continuous Commissioning

- Continuous monitoring with assessments performed at least quarterly
- The building may or may not have been previously commissioned
- Does not adhere to building's original design intent – is concerned instead with trending relative to a baseline and optimizing performance to meet current requirements
- Utilizes building automation system and/or metered energy trend data and/or condition acceptance tests

23

- Develops and implements proposed commissioning measures and
- Measures, verifies, and documents improvements and operational and energy savings.

An important distinction in this form of commissioning is that the process is continuous: steps are taken to maintain the improvements to occupant comfort and safety, operational efficiency, and energy efficiency that have been achieved. The CxA or O&M staff review the system operation and operating and energy trends periodically to identify any problems and to develop improved operation and control schedules. Energy data is reviewed at least quarterly to identify the need for another commissioning tune-up. If building energy consumption has increased, or if the performance efficiency of building equipment and systems has decreased, the CxA or O&M staff performs an evaluation, develops measures to restore the building energy and operational performance, and implements the measures. Ongoing follow-up (at least quarterly) is essential to guarantee the persistence of savings and high levels of energy and operational efficiency over time.

Types of Testing Used

Apart from site observation and interviews with occupants, the bulk of continuous commissioning testing is a combination of analysis of metered and recorded energy data and of condition monitoring. Condition monitoring is one aspect of reliability centered maintenance (RCM) and is common to advanced preventive maintenance programs. It differs from functional performance testing (common to the other types of commissioning) in that it concentrates on the current and predictive *condition* of the equipment, rather than on the *output parameters* and *performance* relative to its design and intent.

Advantages

- Persistence of benefits of the commissioning process is the most obvious advantage to continuous commissioning. The process focuses on finding sustainable engineering solutions based on engineering principles to address problems with building operation, energy efficiency, and/or occupant comfort and safety. An added benefit is a usual decrease in O&M workload and costs.

- Superior operational, energy, and comfort performance is the ultimate goal of continuous

commissioning. The process stresses gathering and analyzing considerable data on occupancy patterns and building operation. Instead of making sure the systems work as designed, continuous commissioning ensures that systems run as efficiently as possible and produce optimal occupant comfort for current conditions. This results in significant savings if the system, as designed, has poor efficiency or a negative impact on occupant comfort.

- Continuous commissioning is proactive and can identify operational problems associated with long-term storage of equipment that are not readily apparent, such as the breakdown of dielectrics, degraded fluids, failed batteries, leaking seals, and flattened bearings.

- Whether the continuous commissioning program is led by a third party CxA or implemented by the facility O&M staff, staff skills will inevitably increase as a result. The O&M staff gains knowledge and skills in diagnosing operating problems and determining and implementing corrective strategies.

- The energy and cost savings resulting from continuous commissioning measures can be used for major systems and equipment upgrades. Continuous commissioning has first costs associated with the training of the O&M staff and the one-time cost of installing a building automation/energy management control system or system metering capability. Once these costs are covered, future savings from operational measures can be applied to the installation of energy conservation measures and other authorized capital improvements. In addition, the continuous commissioning process identifies efficiency measures, reducing the need for additional audits and engineering analysis when programming for major retrofits.

Disadvantages

- Continuous commissioning does not consider design intent – how were the installed equipment and systems intended to operate? Facility uses and occupancy change over time, and it is possible that the design intent is obsolete. It might be beneficial to revise the design intent and use it as a guiding document for O&M; in continuous commissioning, there is no such guide to building operations.

- The installation of a building automation system (BAS), energy management control system (EMCS), or other metering system is required for the monitoring and verification that is essential for tracking the persistence of engineering measures. This can be cost prohibitive to smaller facilities. It can also be a good investment for larger, more energy-

intensive facilities, because a sophisticated energy management or metering system can also be used for load control and other energy management applications.

- Continuous commissioning is most effective, and most cost-effective, when implemented in a facility that already has in place a preventive maintenance program and a highly skilled and trained O&M staff. Lacking this, costs will rise to bring in a qualified CxA to perform the continuous commissioning activities and/or to train the existing O&M staff on continuous commissioning approaches and tests. High O&M staff turnover is also a barrier. However, the cost of training O&M personnel can also be a wise investment, particularly in larger, more complex, and more energy-intensive buildings.

25

References

1. *Continuous Commissioning Guidebook: Maximizing Building Energy Efficiency and Comfort*, Liu, Mingsheng, Ph.D., P.E., et. al., October 2002.

2. *It's Time to Shake Out the Mothballs in Your Mission-Critical Facility*, Soroka, Joe, Comminique (online newsletter of AFCOM), March 2004.

3. *A Practical Guide for Commissioning Existing Buildings*, Haasl, Tudi and Terry Sharp, April 1999.

4. *Retrocommissioning Handbook for Facility Managers*, Portland Energy Conservation, Inc., March 2001.

5. *Whole Building Design Guide*, General Services Administration, available at www.wbdg.org.

Best Practices

■ Carefully consider the short- and long-term plans for your facility. When was the last major equipment change-out? Is existing equipment relatively new, or at the end of its useful life? Is a major renovation on the horizon? These questions will help you decide whether or not your facility can benefit from existing building commissioning.

■ Continuous commissioning adds to your O&M costs, but can be a good investment in large, complex, and energy-intensive buildings.

26

STUDY QUESTIONS

1. What type(s) of commissioning would you consider suitable for your facility?

2. For new construction or major renovation projects, at what phase of the project should the commissioning process ideally start?

3. Under what circumstances would you consider including subject matter experts on the commissioning team?

4. Does commissioning replace or reduce the contractor's Quality Control responsibilities?

5. What are the primary objectives of commissioning?

6. Explain the differences between verification testing, functional performance testing, and condition acceptance testing.

7. Explain the major differences between new building commissioning, retrocommissioning, recommissioning, and continuous commissioning.

8. What is the relationship between any of the forms of commissioning and energy savings?

9. How would you describe the value of commissioning to the Owner? Building occupants and users? The operations and maintenance staff?

Chapter 3
Why Commission?

To better understand the benefits of commissioning, consider how much buildings and their systems have changed over the last couple of decades. Control systems have become highly complex with the migration from pneumatic to direct digital control systems; with more sophisticated building management system (BMS) hardware and software; and with automatic valves, dampers, actuators, and sensors. New technologies have been introduced for life safety and security systems. Buildings must operate with optimal energy efficiency. Indoor air environments have taken on a new importance with regards to mitigation of mold, mildew, and new product emissions as they affect occupant health, comfort, and productivity.

Commissioning also has long-term repercussions on maintainability. Systems may not be installed, adjusted, and integrated to operate optimally. They may be installed with latent manufacturing, transportation, and installation defects. Systems designed and installed with structures that amplify destructive natural harmonics, that get damaged during transport, and that were detrimentally modified on site to "make

it fit" are not uncommon. Equipment literally may self-destruct.

The consequence of most un-commissioned buildings is that the O&M staff inherits systems ripe with problems and inefficiencies. An un-commissioned building may not operate correctly, and without essential O&M information, training, and baseline data, the O&M staff likely cannot respond adequately to occupant complaints. They respond to problem symptoms rather than correcting root causes. Automated systems become bypassed and overridden. Occupants very quickly settle on low expectations and become sensitized to (or very vocal about) the poor building environment, which deteriorates steadily. Energy efficiency suffers, and building performance falls short of the Owner's expectations. These costs are high and well above those for commissioning.

A team from the Lawrence Berkeley National Laboratory, Portland Energy Conservation, Inc., and the Energy Systems Laboratory of Texas A&M University set out to quantify the actual costs and impacts associated with commissioning[1]. *The Cost-Effectiveness of*

"If you always do what you've always done, you'll always get what you've always got; Change makes change."

Anonymous

27

In this Chapter

◆ Commissioning Costs and Return on Investment
◆ Cost Benefits of Commissioning
◆ Barriers and Management Buy-in
◆ Impact on Facility Operations and Maintenance
◆ Impact on Energy Consumption
◆ LEED Certification
◆ Best Practices

Commercial-Buildings Commissioning analyzed results from 224 buildings across 21 states and involved 30.4 million square feet of commissioned floor area (73 percent in existing buildings and 27 percent in new construction). Some results were surprising:

■ Among 85 existing buildings in the study that were being retro-commisioned for the first time, 3,500 deficiencies were found. Approximately 85% of the deficiencies found related to the overall HVAC system. The median cost per building was about $34,000 for commissioning (or $0.27/sf) and resulted in savings of about $45,000 (median) per year or ($0.27/sf/yr). Energy cost savings resulting from the retro-commissioning are estimated to be about 15-percent with a simple payback time of 0.7 years.

■ Deficiencies are expected in older facilities that may have outdated, inefficient equipment

Excerpt from Mills, E., et.al.[1]	Existing Buildings		New Construction	
		Sample Size		Sample Size
No. of Deficiencies Identified	3,500	85	3,305	35
Commissioning Cost ($1,000)	34		74	
Commissioning Cost ($/sf)	0.27	102	1.00	69
Total Savings ($1,000/yr)	45		3	
Total Savings ($/sk/yr)	0.27	100	0.05	33
Whole Building Energy Cost Savings (Median %)	15	74	Not Available	
Simple Payback Standardized to U.S. Energy Prices (Yr)	0.7	59	4.8	35

and systems. But the study also found significant room for improvement in new construction projects. Among 35 new construction projects analyzed, the commissioning process uncovered 3,305 deficiencies. Deficiencies with the air handling and distribution were most common, followed by lighting, and HVAC combined heating and cooling plant. For a median cost of $74,000 per new building (or $1.00/sf), estimated savings were calculated at about $3,000/year (or $0.05/sf/yr) with a simple payback time of about 4.8 years because most new construction projects emphasized a small number of corrective measures rather than a whole-building effort that is characteristic of existing building retro-commissioning. In addition to energy savings, owners in the study reported other benefits such as increased productivity and safety, better indoor air quality and thermal comfort, longer equipment life, and a reduction in change orders and warranty claims.

As these results show, commissioning can be viewed as invaluable to detecting and correcting deficiencies in both new construction/major renovation projects, and in existing buildings. Deficiencies such as design flaws, construction defects, malfunctioning equipment, and deferred maintenance have a host of ramifications, ranging from equipment failure to compromised indoor air quality and comfort to unnecessarily elevated energy use or underperformance of energy strategies. The "newness" of a building does not guarantee fewer deficiencies, as the study demonstrates.

The most frequently cited barrier to widespread use of commissioning is decision-makers' uncertainty about its cost-effectiveness. But because deficiencies are common in both new construction and existing buildings, the bigger financial cost may come from *not* commissioning your building.

COMMISSIONING COSTS AND RETURN ON INVESTMENT

Based on their study, the Lawrence Berkeley National Laboratory team was able to quantify the average cost of commissioning as previously presented:

Retrocommissioning of Existing Buildings

- Cost of commissioning: $0.27/square foot
- Whole-building energy savings: 15 – 20 percent
- Payback time: 0.7 year

29

[1] Mills, E. et al. (2004) *The Cost Effectiveness of Commercial Building Commissioning: A Meta-Analysis of Energy and Non-Energy Impacts in Existing Buildings and New Construction in the United States.* (http:eetd.lbl.gov/emills/pubs/cx-costs-benefits.html)

New Construction/Renovation
Commissioning

- Cost of commissioning: $1.00/ square foot (0.6 percent of total construction costs)
- Whole-building energy savings: N/A
- Payback time: 4.8 year

Whole-building energy savings data is not available for new construction commissioning, as there is no benchmark upon which to measure energy use before commissioning is applied to the project.

The Portland Energy Conservation, Inc. (PECI) studies indicate that on average the cost of operating a commissioned building range from 8 percent to 20-percent below that of a non-commissioned building.

BOMA cost data for office buildings suggest that commissioning can save energy from 20-percent to 50-percent and additional maintenance savings from 15-percent to 35-percent.

The study found that commissioning is cost-effective for both existing buildings and for new construction, across a range of building types, sizes, and energy use. The more complex the building and its systems, the more cost savings commissioning can achieve.

Commissioning costs vary more according to the complexity of the systems, number of pieces of equipment, and objectives or scope of the project rather than by building type. The following graph developed by the Portland Energy Conserva-

30

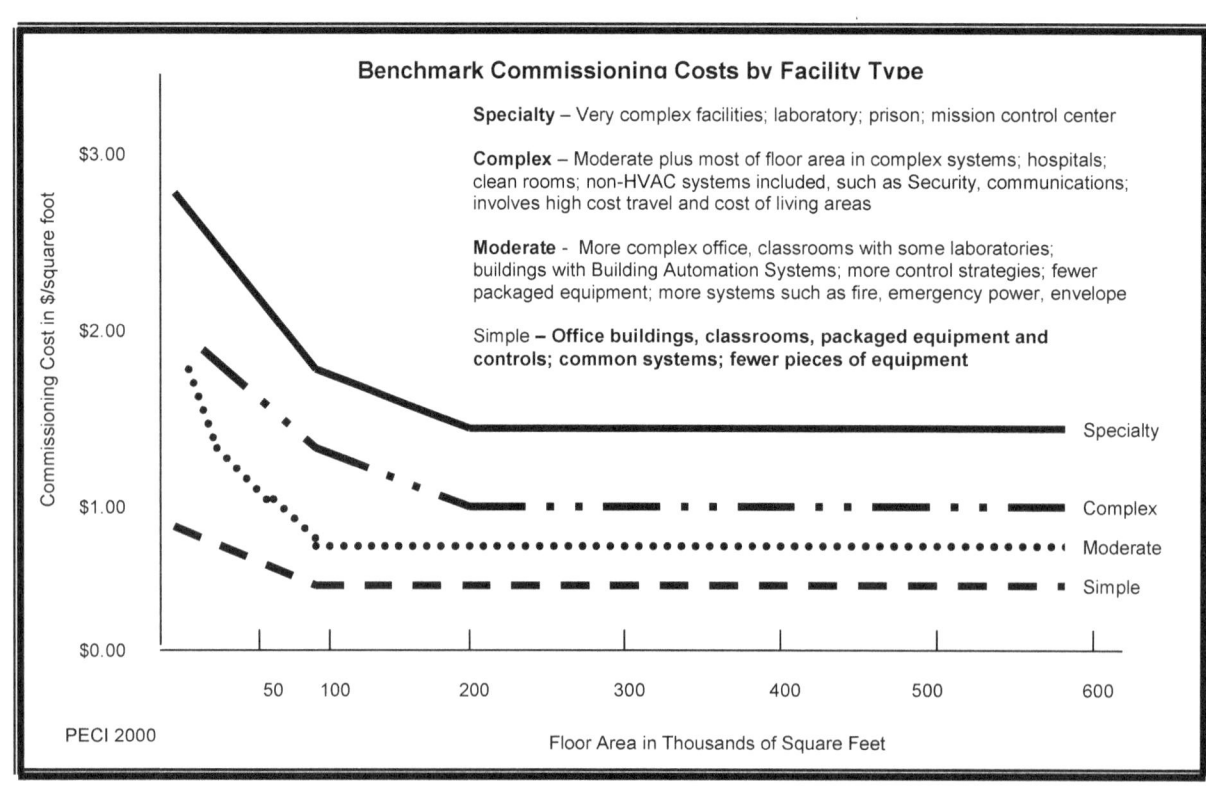

tion, Inc. (PECI) in 2000 illustrates the effects of building size and complexity on the cost of commissioning.

For new construction and major renovation projects, there is typically a three- to five-percent contingency budget for problems that may arise during construction; committing part of this contingency budget to commissioning up front is a smart move to decrease overall costs related to construction deficiencies and to ensure a more efficient building as an end product.

The recognized rule-of-thumb used in the construction industry to estimate return on investment is a $3 savings for each $1 spent on commissioning. (More complex facilities such as laboratories and hospitals, may have greater return on investment ranging from $3 to $11 for each dollar spent on commissioning.

Commissioning provides short- and long-term benefits, so the process should be viewed as an investment rather than an expense. In terms of total cost of ownership, it is important to consider equipment life-cycle costs and energy efficiency in addition to the acquisition or first cost of building equipment and systems.

COST BENEFITS OF COMMISSIONING

The following benefits are common for all types of commissioning:

- **Improved system performance** – Building systems and

technologies are becoming increasingly more complex and energy efficient. But increased system performance will not be realized unless equipment and systems are properly designed, installed, maintained, and optimized to work together in an integrated fashion.

- **Energy savings** – Studies show that commissioned buildings typically save 10 to 20 percent of utility costs compared to similar non-commissioned buildings by working to ensure that system components operate together most efficiently. In particular, properly optimized HVAC and control systems often lead to the greatest energy savings. In contrast, the lack of a commissioning program may lead to under-performance of energy-efficient equipment.

- **Improved thermal comfort** – Commissioning helps ensure thermal comfort. It provides for acceptable levels of temperature and humidity, air movement and ventilation, and the ability for occupants to modify conditions. It provides a better work environment with fewer occupant complaints and enhanced productivity.

- **Extended equipment life and reduced warranty claims** – Commissioning optimizes equipment and systems from day one, meaning fewer

Commissioned buildings typically save 10 to 20 percent of utility costs compared to similar non-commissioned buildings.

31

warranty claims, service calls, reduced energy use, and reduced potential for catastrophic equipment failure. Commissioning ensures that equipment that is properly calibrated, and provides training and documentation to O&M staff that will help achieve extended equipment life.

- **Increased training for building O&M staff** – This is of particular importance given the increasingly complex controls, building management systems, and energy management systems being installed in today's buildings. Inadequate training can lead to sophisticated controls and management systems being shut off, bypassed, and/or not properly programmed and calibrated, reducing the energy savings, safety, and operational efficiencies they were designed to provide.

- **Improved O&M documentation** – Commissioning provides more useful O&M data that is

specific to the systems and equipment installed, details the way the equipment should be operated, outlines preventive maintenance procedures and schedules, and provides information on warranties, spare parts, and vendors. In addition, commissioning and particularly condition acceptance data provide a baseline against which the ensuing maintenance program can be compared and trended.

- **Renewed interest in more closely monitoring facilities maintenance and physical assets** – The commissioning process requires a commitment of internal resources even if the program is outsourced to a commissioning provider. The required program and team building can be a catalyst for an increased interest in facilities improvement and asset management at all levels. Commissioning can also complement an ongoing facilities management process improvement program.

- **Increased occupant comfort, safety, and productivity** – Commissioning addresses common occupant concerns such as thermal comfort, air flow and air quality, and lighting levels to ensure that occupants are comfortable and safe in their work spaces. This can lead to enhanced worker productivity, fewer sick days, and a higher building resale value. Equipment and systems that are installed

---

If a building is designed to use 20-percent outside air to meet IAQ code requirements, less outside air could result in the building being negatively pressurized. This can be conducive to mold growth, cause occupants to become sick, affect productivity, and subject the building to excess energy use. If outside air is greater than the prescribed 20-percent, an excessive amount of energy will be used with consequential higher energy costs. Commissioning can ensure that a building is both efficient and healthy by verifying the functionality of the control system and the responsiveness of each of its associated devices.

Conventional MEP Design	MEP Design with Commissioning
◆ Component driven	◆ Integrated
◆ Focused on compliance with codes and standards	◆ Focused on systems and equipment optimization in addition to codes and standards compliance
◆ Based on first cost	◆ Based on life-cycle cost in addition to first cost

and calibrated properly are also less likely to break down and potentially injure O&M staff.

■ **Improved indoor air quality** – Commissioning ensures that a building is pressurized and has correct fresh air changes for indoor air quality, which decreases mold-related problems and "sick building" syndrome. It can decrease energy costs as well.

■ **Reduced liability** – Any measure that decreases insurance losses contributes to the bottom lines of both the insurer and the insured. Commissioning has been shown by the insurance industry to reduce losses related to fire and wind damage, ice and water damage, power failures, and health and safety. Reduced risk and liability can also increase the asset value of the building.

■ **Reduced O&M costs** – It is much easier and less expensive to maintain a building that operates correctly than to maintain one that does not.

Equipment that has been installed and tested properly and optimized for maximum efficiency will experience fewer problems and requires less unscheduled O&M time. The complete and accurate building documentation that commissioning provides will expedite maintenance troubleshooting. The training provided to O&M staff will increase skill levels and staff effectiveness.

■ **Incentives** – Commissioning has the potential to qualify buildings for utility program rebates and other Federal and State incentives. The Federal Energy Management Program (FEMP) and organizations such as NYSERDA often provide seed money and financial assistance to those Agencies seeking to commission their new or existing building.

■ **Special laboratory pressurization and features** – Laboratories are comprised of many systems and subsystems bound together in complex ways to provide required airflows and

Commissioning provides short- and long-term benefits; the process should be viewed as an investment rather than an expense.

34

pressurization. Many air systems operate around-the-clock resulting in high operating energy costs. Consequently, simple adjustments can yield large savings. More importantly, negative or positive pressurization is used to control the airflow to protect worker health and safety and the environment. Commissioning will check and validate the actual pressure gradients against the design intent documentation as well as the functional performance of interlocking systems to ensure the pressurization is maintained if part of the system fails or is turned off.

The following benefits can be achieved in addition to those listed above for new construction commissioning:

- **Reduced change orders and improved contractor performance and awareness** – Change orders are reduced because many problems and deficiencies are detected through the commissioning process prior to functional performance testing. The process provides a mechanism to correct problems and deficiencies before project turnover, thereby saving Contractor warranty callbacks. Testing and monitoring make contractors more aware of the quality of their work.

- **Improved construction process and project turnover** – Commissioning done properly

provides increased project communication and enhanced conflict resolution. Project turnover includes all functional test forms, O&M and systems documentation, warranty information, and evidence of training activities. Commissioning also provides for follow-up site visits to address any problems that may occur after project turnover.

- **Decreased testing, adjusting, and balancing (TAB) costs** – A reduction in TAB costs can occur because systems and equipment are more likely to be working properly during start-up and verification checks. This allows the TAB contractor to proceed uninterrupted.

It is much easier to quantify the costs associated with commissioning than to track and quantify the benefits. Benefits such as improved energy performance, extended equipment life, improved indoor air quality, and reduced O&M costs, for example, cannot be quantified easily. However, these factors can lead to significant cost if not adequately addressed (which is what commissioning is designed to do):

- Lost productivity during a systems power failure.
- Construction delays due to increased change orders.
- Litigation due to poor indoor air quality, leading occupants to get sick.

The potential cost of each of these examples (and many others) cannot be quantified, but could be very large indeed.

BARRIERS AND MANAGE-MENT BUY-IN

Commissioning is often viewed by building decision makers as an added cost. If commissioning was free of charge, it would most likely be easily adopted by facility owners and operators across the board. But cost is just one barrier to adoption that the commissioning process faces. Common barriers include:

1. The first costs associated with commissioning are viewed by Owners as being relatively high ($0.27/square foot for an existing building; $1.00/square foot for a new construction project). However, there is little quantifiable data in advance on the potential cost savings (both energy and operational) that the commissioning process will generate.

2. There is no *guarantee* of savings – The commissioning process does not necessarily have cost savings as its primary objective. It is designed to optimize all building system and equipment operation to meet the design intent (commissioning) or current building requirements (retrocommis-sioning). The bulk of the actual cost savings achieved through commission-ing is a by-product in the form of *avoided* costs.

3. There is not a sense of "need" for commissioning services. On new construction, quality assurance support is already being provided by the design and construction teams, and in existing facilities, established O&M programs should be able to detect and correct problems without a commissioning process. The Federal Govern-ment's new emphasis on LEED certification for new and existing buildings, which re-quires commissioning, is begin-ning to increase its sense of importance.

4. If a building already went through the commissioning process during new construction or a recent major renovation, there is a feeling of no addi-tional benefit from repeating that process. This attitude fails to realize that buildings, occu-pants, and missions change over time, and any impact on the design intent can impact the equipment's and system's efficiency.

5. The funding source is often a barrier to commissioning new construction. Capital funding, if it does not have a line item for commissioning, is concerned with delivering a finished project

> **Best Practice:** "Give me your worst facility – I will make it work." Tells the Owner that if you can correct *its'* deficiencies and im-prove *its'* efficiency, then you could do wonders with everything else.

on time and within budget. If commissioning is included, it often gets value-engineered out with little regard for the impact on maintenance after final acceptance and turn over. Meanwhile, the O&M and facility staff inherit a building that may be problematic for its lifetime.

Approaches

In addition to outlining the benefits to be achieved through the commissioning process, some approaches for overcoming barriers and obtaining Management buy-in are suggested:

1. **Start with a pilot project**, either a retrocommissioning project with a set of desired goals (i.e., to improve the building indoor air quality), or a commissioning process tied to a planned renovation. A single pilot project will allow changes identified through the commissioning process to be monitored, verified, and its benefits realized and communicated to Management. Consistent commissioning approaches and documentation can then be developed for use on subsequent projects.

2. **Develop a methodology for analyzing the costs and**

So What are the Pitfalls? Some True Experiences:

◆ A university in Washington state contracted for a commissioning consultant but needed to redirect significant resources from an in-house staff to bring the building's systems on line. Facilities management staff found the CxA to be non-responsive and troublesome. They support commissioning to this day, but "not consultants who don't know what they're doing and don't deliver value added to the process."

◆ The contractor for a new 180,000 SF facility in Idaho found a subcontractor adding $6,000 to his bid because of the CxA oversight. The contractor was amazed that the subcontractor added extra money *to get the job done right*! Does that mean that $6,000 can be deducted from his bids if quality is not an issue? If the job is done right the first time, there is no added cost. Commissioning may even increase profits by reducing warranty callbacks.

◆ On a major renovation project, true commissioning was never fully completed because the:

- Owner wanted to move in and start operations

- Contractor wanted to get done and off site at the least cost

- A/E wanted to close the job

- Specifications were weak and

- "Punch List" became a "To Do" list for the maintenance staff

The ultimate effect was that the O&M staff had to divert resources from its preventive maintenance program to correct deficiencies that should have been corrected as part of commissioning acceptance.

benefits of the commissioning process** throughout the life of the project. This methodology should identify and record deficiencies that are found using the commissioning process and estimate the cost avoidance associated with correcting each deficiency.

3. **Keep a thorough record of costs avoided** by identifying problems using the commissioning process. Avoided costs to consider include repair, replacement, installation, design, energy, depreciation, maintenance, revenue loss, and productivity loss. Base cost avoidance estimates on the best data available, and be conservative in how the data is applied . Wildly inflated cost savings work against the goal of establishing a consistent, long-term commissioning program by making savings seem unattainable.

4. **Integrate the commissioning program with the facility's overall energy management program.** Commissioning benefits energy management by ensuring and optimizing the performance of energy efficiency measures and by correcting problems that cause higher than necessary energy use. The energy management program provides benchmarking of energy consumption and demand that makes the impact of commissioning activities easy to quantify during measurement and verification.

5. For existing buildings, **integrate the commissioning program into the facility's overall operations and maintenance program** (continuous commissioning).

6. For existing buildings, **prepare a comparison of facility operating and energy costs with similar buildings** in the area (close to or same age, use, and square footage). A commissioning program may be an easier sell and support if similar buildings have lower operating and energy costs.

7. **Stress the importance of persistent benefits** when addressing the need for recommissioning. Subsequent changes in facility use or tenancy may require a revision in design intent. The addition of increasingly complex energy and building management systems may require additional O&M documentation and training. Setpoints and operating cycles may have been modified by facility O&M staff and/or occupants since the time the facility was last commissioned.

IMPACT OF FACILITY OPERATIONS AND MAINTENANCE

Systems and equipment that are properly installed and calibrated, and fully optimized to work together, will be easier for a facility's O&M staff to maintain. This is one commission-

37

Approaches for Quantifying the Benefits of a Commissioning Program

It is difficult to project in advance what cost savings will result from implementing commissioning in a new construction/renovation project or in an existing facility. There are several methods to employ during the course of the project, however, that will allow you to calculate operational and energy savings when the project is complete. These methods include:

◆ Collect building data and define Key Performance Indicators (KPI) at the outset to pre-define measurement and performance goals.

◆ Establish a metering system in the building to measure energy use. (A metering system has the added benefit of providing an automated exception report if systems are out of tune.) If this is not feasible, use utility bills (previous 12 months) to establish an energy use baseline.

◆ Use an energy simulation program (such as DOE2, Trane TRACE, Carrier HAP, BLAST, Energy Plus) to quickly organize and evaluate results gathered during measurement and verification.

◆ Establish an automated maintenance tracking system that will provide data pre- and post-commissioning on service calls, failure reports, maintenance schedules, etc.

◆ Develop a methodology that tracks the deficiencies that are found using the commissioning process and estimates the cost avoidance associated with correcting each deficiency. Include costs related to repair, replacement, installation, design, energy, depreciation, maintenance, revenue loss, and productivity loss.

◆ Compare maintenance hours, operations cost, and energy use data in the commissioned building or affected area with the same data from a similar, but un-commissioned, facility. In the case of an existing building, compare the same data in the same building or affected area, but before and after the retro- or re-commissioning. (Be careful not to compare costs alone, as cost increases, such as increased energy tariffs, may skew the comparison results.)

Develop an internal report that summarizes cost and benefit data, or have the CxA include the data in the final Commissioning Report. Information gained through these methods will allow you to model potential operational and energy savings on subsequence commissioning efforts. Using this approach will facilitate powerful historical data and allow each project to be evaluated in a value-based manner.

38

A 2004 study conducted by Cornell University correlated worker productivity with indoor environmental conditions. It found that workers produced 74-percent more mistakes and 46-percent less output based on temperature alone when the temperature fell from 77°F to 68 °F. The study estimated that the decreased productivity resulted in a 10-percent increase in labor costs per worker, per hour. "Our ultimate goal is to have much smarter buildings and better environmental control systems in the workplace that will maximize worker comfort and thereby productivity," the professor overseeing the study said. Commissioning by its nature supports that goal and optimizes those very systems. (source: http://ergo.human.cornell.edu/ CUEHEECE_IEQDown.html)

ing benefit that can profoundly impact the O&M staff. Commissioning does require input and effort from the O&M staff. Most of the benchmarking and maintainability issues associated with commissioning are addressed throughout the commissioning process during initial planning, design reviews, verification inspections, functional performance, condition acceptance testing, and training. The O&M staff's effort during commissioning revolves around providing input and feedback as the project progresses, observing the work in progress and system testing, and receiving proper and appropriate training and education on the new and modified installed systems.

Establishing a building performance and O&M program baseline is one of the most important tasks for an O&M organization related to measuring the costs and benefits of a commissioning program. This allows the O&M organization to quantify operational, energy, and other benefits from commissioning, as well as identify unforeseen problems. Most O&M organizations that have adopted common O&M best practices will already have established

such baselines, which include the following performance metrics:

- System Capacity (actual operation compared to full system utilization of plant or equipment)
- Work Orders Generated and Work Orders Closed Out
- Preventive Maintenance Backlog
- Safety Record
- Energy Use
- Inventory Control
- Overtime Logs
- Environmental Record (air and water discharge levels, noncompliance situations)
- Staffing: Absentee and Turnover Rates

Commissioning should produce measurable improvements in each of these performance categories.

A fine-tuned O&M program is one of the keys to achieving persistent commissioning benefits. Successful O&M planning begins early in the commissioning process, often at the design phase, during which O&M staff should participate by providing design recommendation input based on their past hands-on experience.

39

Tracking Energy Efficiency Improvements from Commissioning

It is important to identify and define desirable metrics before collecting energy efficiency improvement data on your commissioning project. The following are sample metrics:

Building Characteristics and Demographics

- Building type (using DOE/CBECS definitions), age, location
- Year building commissioned: new construction/renovation or existing building commissioning
- Reasons for commissioning, deficiencies identified, corrections/improvements recommended

Energy Utilization

- Electricity: kWh/building/year or kWh/sq.ft./year
- Peak electrical power: kW/building or W/sq.ft.
- Fuel: MMBTU/building or kBTU/sq.ft./year
- Purchased thermal energy: MMBTU/building/year or kBTU/sq.ft./year
- Total energy: MMBTU/building/year or kBTU/sq.ft./year
- Energy cost: $/building/year or $/sq.ft./year based on local or standardized energy prices (provide nominal and corrected for inflation) post-commissioning
- Percent energy use savings (total and by fuel) and total energy cost savings post-commissioning
- Persistence index: post-commissioning energy use in a given year/pre-commissioning energy use (unitless ratio)

Commissioning Cost

- $/building or $/sq.ft. (based on nominal costs; can be gross value or net, adjusting for the quantified value of non-energy impacts)
- Commissioning cost ratio for new construction: commissioning cost/total building or renovation construction cost, expressed as a percentage
- Costs for CxA and other parties separately
- Allocation of costs by source of funds (agency capital funds, utility, grant, etc.)
- Total building construction cost (denominator for commissioning cost ratio)

Cost Effectiveness

- Undiscounted payback time (commissioning cost/annualized energy bill savings)
- Normalized to standard energy prices and inflation-corrected to a uniform year's currency levels

Deficiencies and Corrections and/or Improvements

- Number of deficiencies and improvements per building or #/sq.ft.
- Number of combined deficiencies/corrections (per building or per square foot)

Non-Energy Impacts

- Type, quantified when possible: $/building/year or $/sq.ft./year
- One-time or recurring

Adapted from "The Cost-Effectiveness of Commercial-Buildings Commissioning," Mills, Evan et. al., December 15, 2004.

40

A maintenance program should be in place (or be implemented) that includes the following commissioning-related responsibilities:

- Scheduled reviews of operating parameters, schedules, and sequences of operation.
- Scheduled utility bill analysis, or use of energy accounting software, to review for unexpected changes in building energy use.
- Condition monitoring, including vibration analysis, infrared thermography, ultrasonic testing, motor testing, and lube oil analysis, as appropriate, that tracks and trends condition parameters such as equipment alignment and balance, vibration, high-resistance electrical connections, motor condition, fluid leakage, and lubricant condition.
- Tracking of scheduled and unscheduled maintenance for each piece of equipment.
- Periodic reviews of maintenance performance indicators and logs to determine if systems and equipment require tuning.
- Building and O&M documentation updates to reflect changing building requirements and equipment replacement.
- Operator training updated annually.

IMPACT ON ENERGY CONSUMPTION

Building performance today is compromised by a diverse array of physical deficiencies, as found by the Lawrence Berkeley National Laboratory team in their study *The Cost-Effectiveness of Commercial-Buildings Commissioning*. HVAC systems present the most problems, particularly air-distribution systems. In addition, sophisticated energy management control systems installed in today's newer facilities often are not optimized and calibrated properly to deliver the energy savings of which they are capable.

Building commissioning is one of the most cost-effective and far reaching means of improving energy efficiency in buildings. In the Lawrence Berkeley National Laboratory study, a sample of 150 existing buildings found an average whole-building energy savings of 18 percent, with an average payback time of less than

41

one year, when commissioning was applied. A sample of 74 new-construction projects found an average payback time of 4.8 years, although the addition of non-energy impacts can drastically reduce these payback times. There are cost-effective results to be found in a wide range of building types and sizes.

As it relates to energy efficiency goals, commissioning can be seen as a form of risk management. It helps ensure that funds are spent wisely and that the intended energy savings targets are achieved in practice. Commissioning provides a method for defining measurable energy performance targets in the design phase, and for evaluating as-built and as-operated system conditions.

As buildings and energy-efficient technologies become more complex and interconnected, the need for commissioning to ensure optimized energy performance will increase. The Lawrence Berkeley National Laboratory study asserts that the efficacy of energy efficiency programs is directly related to the extent to which they are coupled with commissioning and quality assurance in design and delivery.

LEED Certification

More and more government agencies are demanding "green" buildings that incorporate meaningful strategies for sustainable facility design.

To reduce long-term facility costs and to become more environmentally conscious, many Federal agen-

cies mandate that new and remodeled buildings meet a minimum level of sustainable design certification. One of the most widely adopted programs used by the government to assess building performance and adherence to sustainable design goals is the LEED™ (Leadership in Energy and Environmental Design) certification system. LEED was created by the U.S. Green Building Council (USGBC) and is now recognized and accepted internationally to assess building performance and adherence to sustainable design goals.

The LEED Green Building Rating System is a voluntary, consensus-based national standard for developing high-performance, sustainable buildings. Members of the USGBC, representing all segments of the building industry, developed LEED and continue to contribute to its evolution. LEED standards are available or under development for:

- New construction and major renovation projects (LEED-NC)
- Existing building operations (LEED-EB)
- Commercial interiors projects (LEED-CI)
- Core and shell property (LEED-CS)
- Homes (LEED-H) and
- Neighborhood development (LEED-ND)

To achieve a LEED rating, a whole-building approach must be used. Credits must be earned in several categories, including site selection and configuration, water efficiency, energy, indoor air quality, and sus-

42

Many Federal agencies mandate that new and remodeled buildings meet a minimum level of sustainable design certification. LEED is now recognized and accepted internationally to assess building performance and adherence to sustainable design goals.

tainable building materials. In addition, the USGBC recognizes the effectiveness and benefits to whole-building efficiency that commissioning can achieve, making commissioning a mandatory requirement for LEED certification.

Understanding the LEED certification ratings and how to achieve facility certification with design, construction, and operational credits based on the system has become imperative as government agencies look for ways to become more environmentally friendly, conserve energy, and decrease the operating costs of their real estate.

Commissioning and the LEED approach share a focus on moving beyond a first cost/lowest-cost perspective to consider and account for long-term, life-cycle building expenditures. The LEED approach considers a building's real cost, which includes not only the amount spent to construct the facility but also the ongoing expenses required for building operations and maintenance, which can account for 60 to 85 percent of a building's actual capital expenditures.

BEST PRACTICES

- Consider life cycle costs in addition to first costs when considering commissioning – much of the value achieved by commissioning comes from *avoided* costs rather than quantifying cost savings.

- Integrate a continuous commissioning program with energy

management and preventive maintenance for a whole building approach to operations, energy efficiency, and sustainable design.

- Green buildings that combine energy efficiency, sustainability, and commissioning are not only good for the environment – their impact on occupant comfort, satisfaction, and safety can result in serious savings for the facility operator.

- Resist value engineering commissioning out of projects because of cost. The cost of commissioning is such a small percentage of the overall project cost that it's removal is unlikely to swing a project back into budget. More importantly, it is wrong to compromise quality as a result of budget concerns.

43

References

1. "The Cost-Effectiveness of Commercial-Buildings Commissioning," Mills, Evan, et. al., Lawrence Berkeley National Laboratory, December 15, 2004.

2. "Building Commissioning: The Key to Quality Assurance, "U.S. Department of Energy.

3. "Persistence of Benefits from New Building Commissioning," Friedman, Hannah, et. al., August 2002.

4. "Development of an Integrated Commissioning Strategy Cost Model," Veltri, Anthony, National Conference on Building Commissioning, May 8-10, 2002.

5. "Quantifying the Cost Benefits of Commissioning," Altwies, Joy E. and Ian B.D. McIntosh, National Conference on Building Commissioning, May 9-11, 2001.

6. "No Bugs: Use Commissioning to Make Sure Buildings Work," Coyne, Mike, Nashville Business Journal, December 13, 2004.

STUDY QUESTIONS

1. What arguments would you offer to convince the owner that commissioning is the right thing for his facility?

2. What are the standard barriers against the commissioning process?

3. What are ten benefits derived from commissioning and explain how they are beneficial?

4. Explain how commissioning reduces the number of change orders in new construction?

5. How does commissioning contribute to new construction acceptance and project turnover?

6. How can commissioning reduce the cost of TAB?

7. How can the O&M program optimize the performance of your facility?

8. How does the LEED certification impact the quality of life of the building occupants?

Chapter 4
Commissioning Management

An emphasis on team building and teamwork will lead to greater success from the commissioning process. It fosters a positive approach to commissioning activities, rather than the attitude that the commissioning process creates more work or encourages "looking" for problems that do not exist. Effective commissioning management encourages team members to work together to identify problems (existing and potential) and help correct these problems before they grow.

There are two teams that will be discussed in this chapter: the internal management team that must be in place and prepared to support the commissioning process, and the commissioning team that is assembled when a project is underway.

DEVELOPING COMMISSIONING STRUCTURE AND MANAGEMENT SUPPORT

A commissioning program has little chance of achieving measurable success unless its goals and objectives are clear and the program has the full support of your internal management organization.

There are three steps to developing a strong internal structure to support commissioning: determining the need for commissioning, obtaining the support of various building stakeholders, and defining an internal commissioning team.

Step 1 - Determine the Need for Commissioning

Commissioning is becoming more and more common, but its value should still be carefully evaluated and management support for commissioning obtained. For new construction and renovation projects, commissioning will produce the biggest payoff if the facility and its systems/equipment are very complex (laboratories, telecommunications, medical), and if the facility has a very large square footage. Commissioning will not produce the same level of savings for simpler, smaller facilities.

In the case of retrocommissioning, there are several factors to consider:

■ Are equipment and systems programmed to be replaced

45

"What you always do before you make a decision is consult. The best public policy is made when you are listening to people who are going to be impacted. Then, once policy is determined, you call on them to help you sell it."

Elizabeth Dole

46

within a year or two? Hold off on retrocommissioning until after the change-outs are made.

■ Are equipment and systems continually failing, and major system design problems seem to be the culprit? Forgo retrocommissioning and focus on making commissioning a part of a redesign effort.

■ Are equipment and systems outdated but not necessarily broken (and not near the end of their useful life)? Retrocommissioning can tune up an old system in a cost-effective manner.

■ Are the equipment and building systems relatively new but subject to periodic failure and not operating efficiently? Retrocommissioning can identify and prioritize needed equipment repairs and improvements based on their potential return on investment.

Step 2 - Obtain Support

Federal facility and O&M managers must obtain full support from their management structure to implement a successful commissioning program. One way to gain the approval and support of Management is to develop a written statement of commissioning objectives, goals, costs, and benefits. Approach commissioning by equating it with increased productivity, energy efficiency, safety, and occupant satisfaction.

Commissioning is easier to integrate into a new construction or major renovation project than in existing buildings. New projects involve substantial capital expenditure, usually from a separate capital investment budget, and Management will be interested in methods of guaranteeing quality assurance and getting the best building for their investment. Also, the new construction or major renovation commissioning process does not involve intensive work on the part of the Owner or facility O&M staff beyond their involvement in the project itself.

Retrocommissioning is more difficult to sell to Management. Persistent building equipment and system failures, or energy and operational efficiency losses are perceived to fall under the scope of the existing

O&M program ("Why can't maintenance just fix the problem?"). The following benefits can be outlined to counter this:

- **Asset Management –** retrocommissioning increases the ability of the O&M staff to provide quality services to the building's occupants, and the building's net operating income increases when it is operated as efficiently as possible.

- **Risk Reduction –** retrocommissioning identifies equipment and system deficiencies that could lead to tenant loss, decreased occupant productivity, reduced equipment life, reduced indoor air quality, "sick building" syndrome, unhealthy effects of mold and mildew growth, and higher utility bills.

- **Internal Benchmarking –** retrocommissioning provides a benchmarking tool for building operational performance, allowing an ongoing record to be kept of quality control, and for condition baselining to be used to measure and ensure maintenance performance.

- **Energy Management –** retrocommissioning provides a low-cost method for obtaining energy efficiency savings without capital outlay (for instance, a chiller may not need to be replaced, but rather its controls recalibrated to optimize its performance). Retrocommissioning can increase energy

efficiency in buildings by as much as 15 percent.

- **Low First Costs –** retrocommissioning is typically a one-time event, and does not necessarily involve all building systems and equipment. The retrocommissioning process involves evaluating building performance and choosing the most high-priority (least efficient) systems and equipment upon which to focus attention.

Another stakeholder vital to implementing a successful commissioning program is the facility O&M staff. Commissioning for new construction and major renovation is again easier to sell to the O&M staff, as they will most likely appreciate thorough training and O&M documentation on newly installed equipment. For retrocommissioning, it is important to stress that the process is there to make their jobs easier, not harder.

If commissioning duties are handed over to the O&M organization, make sure to stress that these activities can and should be incorporated into their regular preventive maintenance program as part of continuous commissioning. If a third party commissioning provider is hired to perform retrocommissioning, assure the O&M staff that the process will identify equipment, systems, and approaches that are not working as well as they should.

The end result will be a better building that is easier to maintain, with less trouble calls and more time to proactively implement preventive

> ### Benefits of Retrocommissioning:
>
> Asset Management
> Risk Reduction
> Internal Benchmarking
> Energy Management
> Low First Costs

47

maintenance tasks. Make the O&M staff a partner in the commissioning process.

Finally, the needs and desires of the building's occupants must be considered. Conducting a survey of occupant satisfaction is a good place to start (and can provide information for Management on areas of dissatisfaction that may justify performing retrocommissioning). Building occupants will support a retrocommissioning program if it will lead to better thermal comfort, air quality, and lighting levels.

Step 3 - Define an Internal Commissioning Team

When the need for commissioning has been recognized and accepted by Management, O&M, and occupants, define a team that will man-

age the process for the Owner.

For a new construction or major renovation project, this team will include the facility management staff person who is acting as the Owner's project manager for the project. If this individual does not want to take on the responsibility of monitoring the commissioning aspects of the project, another person on the facility management staff should be appointed.

It is vital that each party involved in the project have access to a single point of contact for commissioning issues who is also the Owner's representative. In addition, a lead contact should be appointed from the facility O&M organization to coordinate O&M activities related to commissioning.

For a retrocommissioning project, representatives from facility man-

Tip: Build a Building Performance Team

Many Federal agencies have voluntary energy management teams at their facilities: groups representing different stakeholders that meet and share ideas on improving energy efficiency. Consider expanding this idea to overall building performance.

Solicit volunteers from Management, O&M, and tenants to meet periodically to discuss problems (such as comfort and safety) and opportunities to address these problems while increasing the building's energy and operational efficiency. To motivate participants, consider applying for LEED™-EB or ENERGY STAR Label for Building certification: both programs incorporate commissioning and provide public recognition for efficiency improvements.

agement and O&M should develop an organization and define duties based on the size and scope of the project. If a third party CxA is hired, facility management and O&M should each provide the CxA with a single point of contact.

COMMISSIONING AUTHORITY OPTIONS

Put most simply the Commissioning Authority (CxA; also sometimes called Commissioning Agent) is the designated person or company that plans, coordinates, and oversees the commissioning process. This person or company directs the day-to-day commissioning activities of the project. The CxA does not have a direct oversight role, like the construction manager, but rather informs installing contractors, the construction manager, and the Owner of observed deficiencies.

There are several options for obtaining the services of a CxA for a project:

Independent Third Party

Most appropriate for:
◆ New construction or major renovation projects, retrocommissioning, and recommissioning of all building types and system complexities

An independent third party CxA is the most common option for providing commissioning services today, and the one most often utilized and recommended by Federal agencies. This person or firm is hired by the

Owner and offers the most objective perspective of any of the other CxA options described in this section.

For large or complex projects, and in buildings with highly integrated and sophisticated systems, potential savings resulting from objective commissioning will likely outweigh the cost of employing an independent third party commissioning provider. And in existing buildings, an independent third party CxA brings a new perspective to the building, has no investment in existing maintenance approaches, and therefore may be more likely to find additional opportunities for improvements and savings.

An independent third party provides an autonomous and independent judge of quality with minimum possible conflicts of interest.

Mechanical or Electrical Contractor

Most appropriate for:
◆ Retrocommissioning of specific systems (e.g., HVAC or electrical systems)
◆ New construction or renovation involving less than 20,000 square feet

Mechanical and electrical contracting firms may already perform comprehensive performance tests and diagnostic procedures for equipment and systems they install. Expanding the scope of work of the mechanical or electrical contractor to include commissioning is an alternative when the project is small

49

An independent third party provides an autonomous and independent judge of quality with minimum possible conflicts of interest.

50

and the requirements of commissioning already detailed clearly in the project specifications.

There are significant drawbacks to consider, though. Contractors may have the knowledge to test the equipment they install, but they may not have experience in testing or diagnosing system integration problems. Further, conflicts of interest may arise from a mechanical or electrical contractor appraising his own work, as identifying deficiencies found through commissioning may increase the contractor's project costs.

Another closely related option, with the same advantages and drawbacks, is an independent division or subsidiary of the construction manager, mechanical contractor, or electrical contractor, an option that may be available from a substantially large contractor.

Design Professional

Most appropriate for:
◆ New construction or major renovation commissioning involving more than 20,000 square feet, with complex design considerations

A design firm (architect/engineer, or A/E) is typically already on board when considering commissioning for a large, complicated new construction or major renovation project.

The advantage of this option is that the A/E is already familiar with the design intent of the project, which somewhat reduces the costs of management. This presents a

greater cost advantage as the complexity and size of the project grows. Commissioning costs are not generally included in the A/E's professional fee, so the scope of work and bid specifications for the project design team will need to be expanded to include commissioning to ensure that the associated costs are captured in the A/E's bid.

The biggest drawback to using the A/E as the CxA is that the firm may not have adequate experience in day-to-day construction processes, troubleshooting, and systems testing required to effectively provide commissioning services. Also, conflicts of interest may arise during the design phase, as identifying potential design problems may increase the A/E's project costs. However, the design professional has a fiduciary relationship to the Owner and is legally bound to act in the Owner's best interests.

In-house Facility Personnel

Most appropriate for:
◆ Recommissioning and continuous commissioning

Choosing to have in-house facility management and O&M personnel assume commissioning responsibilities can be cost-effective and have persistent results, particularly if your facility already has progressive preventive maintenance and quality assurance programs in place (continuous commissioning can be integrated easily into existing maintenance programs, for instance). In-house staff knowledge of equipment, systems, controls, operating strategies, and maintenance procedures

---

Are You Considering Bringing Commissioning In-House?

Most facility owners and managers find that the benefits of employing a highly qualified third party commissioning authority outweigh the costs.

But in the case of continuous commissioning, recommissioning, and some retrocommissioning projects, it may be most economically feasible to hire or appoint a Commissioning Manager to work in-house. Assigning or hiring this role sends a message to the O&M staff that the commissioning program being undertaken is a high priority for the organization, and gives facility management a go-to person to monitor and track progress resulting from the commissioning program.

The Commissioning Manager hired or appointed must have a background in commissioning and condition acceptance testing, and should have the skill and desire to develop and carry out all aspects of the commissioning program.

If hiring or appointing a Commissioning Manager is cost prohibitive, consider hiring a qualified third party CxA to develop and deliver a training program on the commissioning process and systems testing to your O&M staff.

---

can minimize costs. The O&M staff may already perform many of the tests required by recommissioning and continuous commissioning.

Familiarity with their own building can be a serious drawback to using in-house facility management and O&M staff to provide commissioning services, however. In-house staff may be too "close" to how the building currently operates to be able to test and evaluate with full objectivity. Additional training on testing for system integration may be required, as well as training on the full operational potential of sophisticated building automation and energy management control systems. Finally, there should be some incentive offered for assigning commissioning activities beyond regular maintenance duties; the staff must be experienced, motivated, and

available before commissioning should be considered.

CxA Qualifications

Qualifications for the CxA will vary depending on the type of commissioning and size and scope of the project. At a minimum, the CxA (which can be an individual or a company) should possess knowledge and experience with developing commissioning test plans and directly coordinating and overseeing the commissioning process *in practice*.

The CxA qualifications presented in this section should be modified to fit each particular project, but provide a good start for evaluating CxA candidates.

CxA Qualifications

The individual or firm should demonstrate knowledge and experience in the following areas:

1. Designing, specifying, and/or installing building HVAC and mechanical control systems.
2. Operation and troubleshooting of HVAC systems, energy management control systems, and lighting controls systems, including field experience.
3. Controls systems, control sequences, and integrated operations.
4. Performing condition acceptance testing to detect latent manufacturing, transportation, and installation defects.
5. Writing functional performance test plans.
6. Designing energy-efficient equipment and systems and optimizing control strategies.
7. Providing building operation and maintenance and O&M training.
8. Testing and balancing of both air and water systems.
9. Monitoring and analyzing system operation using energy management control system trending and stand-alone data logging equipment.
10. Testing instrumentation.
11. Inspecting and testing electrical power distribution and generation equipment and systems.
12. Developing quality processes and preventive maintenance approaches.
13. Familiarity with LEED point criteria and requirements.

The individual or firm may also meet the following optional requirements:

1. Excellent verbal and writing communication skills.
2. Highly organized and able to work effectively with the building management, design team, installing contractors, and the O&M staff.
3. Education and professional registration – the individual, or in the case of a company the lead individual proposed by the firm, should have bachelor's degree in an applicable area (mechanical engineering, electrical engineering, etc.), as well as either Professional Engineer licensure or other technical training and past commissioning and field experience.
4. Depth of experienced, qualified personnel, and capability to sustain loss of assigned personnel without compromising quality and timeliness of performance.
5. Status as an independent contractor – the individual or firm chosen to provide commissioning services should not be an employee or subcontractor of the general contractor, construction manager, design team, or any other contractor on the project.

The *Past Experience Evaluation* form provided on the following page provides a helpful guide to evaluating the experience and qualifications of the CxA candidate. Also be sure to ask for specific project experi-

52

Past Experience Evaluation

The following questions should be asked when evaluating the CxA candidate's past experience providing commissioning services (and should be modified to fit each particular project). These questions can apply to individuals or companies:

1. What is the overall percentage of your business that is devoted to providing commissioning services?

2. How long have you offered commissioning services?

3. How many commissioning projects have you performed in the last five years? How many commissioning projects similar in size and scope to the project being procured have you performed in the last five years?

4. If applicable, how many LEED projects have you commissioned in the last five years?

5. Are you a registered engineer (individual)? How many registered engineers on staff have directed commissioning projects (company)?

6. How many engineers on staff have performed commissioning projects (company)?

7. How many technicians on staff have performed commissioning projects (company)?

8. Indicate your experience in the following:

❏ Package or split HVAC	❏ Building envelope
❏ Chilled water systems	❏ Fire and life safety
❏ Heating water systems	❏ Plumbing
❏ Building automation systems (BAS)	❏ Elevators
❏ Variable frequency drives (VFD)	❏ Compressed air systems
❏ Lighting controls	❏ Data and communications
❏ Daylighting	❏ Steam systems
❏ Electrical	❏ Clean rooms
❏ Occupancy sensor (lighting controls)	❏ Steam rooms
❏ Energy power generation	❏ Other: _____

53

ence information if your project involves complex system requirements, such as those found in laboratories, medical facilities, telecommunications, computer rooms, etc.

ROLE AND RESPONSIBILITIES OF THE CxA

It is the CxA's responsibility to work with all team members to accomplish the goals set forth in the commissioning plan. The CxA is not responsible for design concept, design criteria, compliance with codes, design, or general construction scheduling, cost estimating, or construction management. The CxA may assist with problem-solving deficiencies or non-conformance, but ultimately that responsibility resides with Owner and the construction manager/general contractor.

The primary role of the CxA is to develop and coordinate the execution of a testing plan, and to observe and document performance to verify that systems are functioning in accordance with the documented design and the contract documents. The installing contractors provide all tools or the use of tools to start, verify, and functionally test equipment and systems, except for specified testing with portable data-loggers that are supplied by the CxA.

Although the scope of the CxA's role varies for different types of projects (new construction commissioning, retrocommissioning, etc.),

the CxA's primary tasks generally include the following:

- Developing a commissioning plan. This includes a preliminary commissioning schedule for inclusion with the bid documents and coordinated with the construction schedule.

- Preparing commissioning specifications that identify the roles and responsibilities of all contractors and project team members. This includes helping the Owner's representative incorporate commissioning specifications in the bid documents.

- Reviewing design and construction documents, drawings, and submittals for commissioning and O&M considerations, as well as for energy performance, water performance, maintainability, sustainability, indoor environmental quality, and environmental impacts. This includes providing comments to the Owner and A/E and conducting follow-up meetings to ensure that comments are addressed.

- For retrocommissioning projects, performing an assessment of operation and maintenance procedures and energy use for existing systems. This includes providing an assessment of various systems' operational costs relative to maintenance and utilities, including equipment life cycle and project energy savings.

54

■ Conducting commissioning team meetings and providing meeting minutes, including project coordination and contractor schedules. This includes ensuring that other commissioning team members understand their specified commissioning responsibilities.

■ Establishing project communication and documentation controls and protocols relative to commissioning.

■ Identifying and documenting contractor and vendor responsibilities as defined by the contract documents, including commissioning criteria, procedures, report formats, functional performance testing and acceptance criteria, benchmarking installations, documentation, and training requirements. This includes notifying commissioning team members of these requirements.

■ Developing diagnostic test plans, writing verification and functional performance test

Commissioning Documentation

The requirements of commissioning documentation vary widely depending on the type of commissioning project. Documentation that is vital to the commissioning process, and should be required by the owner depending on project scope, include:

◆ Pre-construction deficiency list and impact

◆ Commissioning plan, updated as the project progresses

◆ Complete commissioning specification describing commissioning activities and roles and responsibilities of all parties

◆ Current and updated commissioning schedules

◆ Test forms and report formats

◆ Final commissioning reports

◆ Equipment condition baseline data for on-going maintenance

◆ Contractor performance evaluation reports

◆ Documentation pertaining to benchmarking, testing, and training

◆ Deficiency reports, updated weekly

◆ Equipment and systems O&M cost report identifying impact to the project and facility

◆ Updated as-built drawings

◆ Systems manuals

◆ Operation and maintenance manuals

forms, and establishing testing schedules and sequences.

- Overseeing and witnessing start-up, verification, and functional performance tests; verifying test results; consulting on problem resolution; providing recommendations and dispute resolution; and recommending acceptance to the Owner.

- Assisting the Owner and A/E in finding and achieving the requisite points needed for LEED certification, as applicable.

- Coordinating and overseeing all testing and balancing (TAB) and duct pressure testing. This includes reviewing and approving preliminary and final TAB reports.

- Tracking and reporting project progress, including all deficiencies, deviations, change orders, and maintainability issues. This includes monitoring A/E responses to requests for information (RFI) and change orders for potential impact on system operation and maintenance, as well as tracking and reporting on commissioning punch list items.

These items should be compiled and submitted as periodic commissioning progress reports to the Owner's representative and the project construction manager.

- Reviewing operation and maintenance manuals; ensuring that they adequately consolidate all O&M, as-built, warranty, and commissioning data.

- Identifying training requirements and organizing, coordinating, and participating in O&M personnel training.

- Writing a final commissioning report to document the evaluation of the system's capabilities with respect to the Owner's needs and the documented design intent.

- Coordinating and supervising required seasonal and deferred testing and deficiency corrections.

- Performing a site visit 10 months into the 12-month warranty period to review with O&M staff the current building operation and the status of outstanding issues related to the

56

original and seasonal commissioning. This includes:

- Interviewing facility staff and identify problems or concerns they have operating the building as originally intended.

- Making suggestions for improvements and recording any changes into the O&M manuals.

- Identifying areas that may come under warranty or under the original construction contract.

- Assisting the facility staff to develop reports, documents and requests for services to remedy any outstanding problems.

■ If requested, assisting in the development of a preventative maintenance plan, a detailed operating plan, an energy and resource management plan, and/or as-built documentation.

ROLES AND RESPONSIBILITIES OF OTHER COMMISSIONING TEAM MEMBERS

The CxA is responsible for the success of the commissioning program, but the formation of a committed commissioning team is critical for integrating the commissioning process into the project at hand. Each member of the commissioning team

has important roles to play that help determine the success of the commissioning program.

All parties of the commissioning team are responsible for ensuring that equipment and systems are installed in a quality manner and that any problems are identified as early as possible. Each individual worker has the authority and responsibility to identify poor quality workmanship and to recommend stopping work if serious problems are discovered.

A commissioning team differs from the construction team, as it is made up of representatives from various organizations and trades that serve as commissioning points of contact. The participants will vary: a retro-commissioning project may involve only the Owner and/or facility manager and the facility O&M staff, while a new construction or major renovation commissioning team will expand to include the design team (A/E), construction manager or general project manager, installation contractors, controls contractor, and TAB contractor.

A commissioning scoping meeting is the first step to establishing the commissioning team. The CxA, acting as a representative for the Owner (and with the Owner's input) will use this meeting to describe each commissioning team members' roles and responsibilities. The commissioning process is described, and the schedule for commissioning activities is presented.

The CxA should review the scope of the project and advise the Owner

57

on how roles may be modified according to the complexity and size of the project. Therefore, the roles and responsibilities presented in this section are general suggestions, and should be modified to fit the unique requirements of each commissioning project.

Owner/Facility Manager

The Owner, or the facility manager acting as the Owner's representative, is responsible for appointing the CxA, developing goals for the commissioning project, and communicating these goals to the other commissioning team members. Specific commissioning responsibilities include:

- Hiring or appointing the CxA and other members of the project team.
- Determining and clearly communicating to the A/E and CxA the building project expectations, objectives, and focus.
- Working with the CxA to define the goals of the commissioning program.
- Determining the project budget, schedule, and operating requirements.
- Assigning commissioning roles and responsibilities to the in-house O&M staff.
- Facilitating communication between the CxA and other project team members.
- Approving verification and functional performance tests upon completion (and as recommended by the CxA).
- Attending O&M training sessions, as appropriate.

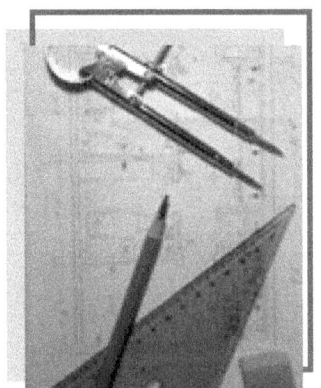

All parties are responsible for ensuring that equipment and systems are installed in a quality manner and that problems are identified as early as possible.

A/E Design Team

An architect/engineer (A/E) design team is normally not used for existing building commissioning projects, but is a vital team member in new construction and major renovation projects. Specific commissioning responsibilities include:

- Documenting the Owner's design intent.
- Including commissioning activities in the bid specifications.
- Providing any design narrative and sequences documentation requested by the CxA.
- Clarifying (along with the installing contractors) the operation and control of commissioned equipment in areas where the specifications, control drawings, or equipment documentation are not sufficient for writing detailed testing procedures.
- Revising project design, if necessary, based on input from CxA design review on commissioning and O&M considerations, energy performance, water performance, LEED certification, maintainability, sustainability, indoor environmental quality, and environmental impacts.
- Monitoring construction activities and informing commissioning team members of change orders and A/E responses to RFIs.
- Documenting any new systems and controls.
- Reviewing and approving project documentation, including shop drawings, operation and

maintenance manuals, submittals, and as-built drawings.

■ Reviewing the commissioning plan and functional performance test plans for design issues.

Construction Manager/General Contractor

The construction manager (CM) or general contractor (GC) is hired by the Owner to manage and construct the project. Like the A/E, the CM/GC is not used normally for existing building commissioning projects, but is a vital team member in new construction and major renovation projects. The CxA represents the owner, but on the project communicates directly with the CM/GC. Specific commissioning responsibilities include:

■ Coordinating and facilitating interaction between the commissioning team and other project team members.

■ Providing a copy of all construction documents, addenda, change orders, approved submittals, and shop drawings related to commissioned equipment and systems to the CxA.

■ Including written requirements for submittal data, O&M data, commissioning activities, and training when hiring installing contractors, and ensuring that each installing contractor meets these requirements as the project progresses.

■ Working with the CxA and installing contractors to coordinate and schedule verification,

functional performance, and conditioning acceptance tests.

■ Working with the CxA to compile and organize O&M documentation and coordinate O&M training sessions.

■ Working with the CxA to prepare O&M manuals, including clarifying and updating the original sequences of operation to as-built conditions.

■ Ensuring that installing contractors execute seasonal and deferred functional performance testing, and that any necessary adjustments are made to the O&M manuals and as-built drawings for applicable issues identified in during seasonal and deferred testing.

Installing Contractors

Installing contractors (mechanical, electrical, controls, HVAC, fire protection, etc.) have the following commissioning responsibilities:

■ Assisting with the development of functional performance tests.

■ Along with the A/E, clarifying the operation and control of commissioned equipment in areas where the specifications, control drawings, or equipment documentation are not sufficient for writing detailed testing procedures.

■ Using test plans and forms provided by the CxA for equipment and system start-up, verification, functional performance, and condition acceptance testing (performed by the installing contractor and witnessed by the CxA).

Condition acceptance testing includes vibration analysis, alignment, balance, infrared thermography, oil analysis, etc.

59

- Participating in the resolution of equipment or system deficiencies, non-compliance, and/or non-conformance identified during commissioning; providing re-testing services if needed.
- Attending commissioning team meetings and providing input into the commissioning schedule.
- Adjusting building systems and documenting system start-up.
- Providing complete operation and maintenance manuals for installed equipment as soon as all of the submittal documentation has been approved (at or about 60 percent project completion).
- Providing training for the building O&M staff (or coordinating training by the manufacturer's representative).
- Participating in seasonal and/or deferred testing.

Equipment and Systems Suppliers/Manufacturers

The suppliers and manufacturers provide specified systems, components and equipment to the contractor and subcontractors. To facilitate commissioning, suppliers and manufacturers should be asked to conduct factory and site performance tests and provide O&M documentation and training for specific equipment. Specific commissioning responsibilities include:

- Providing all requested submittal data, including detailed start-up procedures and specific responsibilities, to keep warranties in force.
- Assisting with equipment testing if required by (and per agreement with) installing contractors.
- Providing any special tools and instruments (only available from vendor, specific to a piece of equipment) required for testing.

The Contributions of Building Occupants and Users Are Invaluable

On a project to commission a building at a U.S. Armed Forces service academy, representatives from the intended building occupants and users participated in a series of commissioning meetings during the early project planning and design stages.

They provided valuable information and insight that could have otherwise required costly change orders or post-project modifications. Examples include the courtroom staff identifying the mandated location of an evidence vault, which required special flooring structural reinforcement and extraordinary security modifications; occupants pointing out that the planned dumpster pad/loading dock location is below and downwind of the building's proposed conference area and ventilation outside air intake; and occupants and users identifying preferred locations for copy machines because of noise and kitchen/microwave oven areas because of odors, noise, and accessibility.

60

- Providing information requested by the CxA regarding equipment sequence of operation and testing procedures.
- Reviewing test procedures for equipment installed by factory representatives.
- Providing detailed O&M manuals and related information specific to the installed equipment and systems as soon as all of the submitted documentation has been approved (at or about 60 percent project completion).

Facility Operations and Maintenance Staff

The building O&M staff provide continual services to effectively operate and maintain building systems, subsystems, and equipment. Specific commissioning responsibilities include:

- Assisting where required in defining and reviewing maintainability requirements in the design intent and defining training requirements in the commissioning plan and specifications.
- Observing and assisting with as much of the functional performance and condition acceptance testing as possible, and using the baseline data collected to establish or refine the maintenance program.
- Attending all O&M training sessions.

Building Occupants and Users

A representative of each building occupant and user organization should participate in early project planning meetings (as the design intent and early design layouts are developed) to identify special requirements and concerns to the owner.

BEST PRACTICES IN COMMISSIONING MANAGEMENT

- Build the team to succeed – the importance of a well-defined internal commissioning management team *and* an integrated project commissioning team cannot be overstated.

- Make facility management staff, O&M staff, and building occupants stakeholders in building performance, and solicit their input into the commissioning process.

- Set commissioning goals with input from your internal commissioning management team in areas of operations and maintenance, safety and comfort, energy efficiency, sustainable design, and LEED certifications.

- Make careful evaluation of your new construction project, renovation project, or existing building to determine the appropriate level of commissioning for the various systems and equipment.

- In almost all cases, hire a qualified, independent third party CxA through a competi-

61

tive bid process. Consider carefully their approach, qualifications, and what they have to offer in terms of expertise, inclusions, and deliverables. Carefully weigh the drawbacks if choosing to obtain commissioning services through a different option.

■ If assigning any commissioning duties to the O&M staff, or instructing the O&M staff to add continuous commissioning and/or condition monitoring elements to the O&M program, make sure that they receive thorough training on commissioning processes and systems testing procedures.

■ Facilitate the integration of commissioning into the normal design and construction process for new construction or renovation projects in order to minimize potential scheduling conflicts and time delays.

■ Foster teamwork and cooperation from all commissioning team members to minimize adversarial relationships and maximize teamwork.

References

1. "Building Commissioning: The Key to Quality Assurance," U.S. Department of Energy.
2. "A Practical Guide for Commissioning Existing Buildings," Haasl, Tudi and Terry Sharp, April 1999.
3. "O&M Best Practices: A Guide to Achieving Operational Efficiency," Sullivan, G.P. et. al., May 2002.
4. "15 O&M Best Practices for Energy-Efficient Buildings," Portland Energy Conservation, Inc. December 1999.Best Practices in Commissioning Management.

STUDY QUESTIONS

1. Occupants in an existing 20-year old, 100,000 square foot building are complaining about poor indoor air quality conditions (temperature and humidity). Private space heaters are common during the winter, mold is apparent on some exterior walls, and documents occasionally get damaged from moisture dripping from duct systems. These problems have been pretty continuous for at least a year and little seems to be done, in the occupants' eyes, to mitigate the situation. What are some situations where retrocommissioning would *not* be appropriate?

2. Recognizing that commissioning adds up-front cost to the project, what do O&M managers have to do to gain the support of their management organization to commission their facilities?

3. Why is retrocommissioning of existing buildings commonly more difficult to sell to Management than is commissioning of new buildings?

4. What are five points that can be made to management to counteract the "Why can't maintenance just fix the problem?" mentality?

5. What are the qualifications a Commissioning Authority should possess?

6. Discuss the questions you would ask a Commissioning Authority candidate.

7. What are the benefits and drawbacks of performing commissioning services in-house with your existing operations and maintenance team acting as Commissioning Authority?

8. What are the aspects of a new construction or renovation project for which the CxA is not responsible?

9. How does the commissioning team differ from the construction team?

10. Define the role of the Owner or facility manager in the commissioning process.

11. On your new building commissioning project, who would *you* include as your commissioning team members, and why?

12. Who would *you* have on the team to retrocommission an existing building system(s), and why?

64

This page left blank intentionally

Chapter 5
Commissioning Process

The overall responsibility of the Commissioning Authority (CxA) during any and all phases of a commissioning project is to coordinate and direct the commissioning activities in a logical, sequential, and efficient manner. This is done using consistent protocols and forms, centralized documentation, clear and regular communications, and consultations with all involved parties.

The commissioning approaches described in this guide are as different as they are similar. This chapter breaks each commissioning approach down by phase.

As used in this chapter, *Owner* can refer to the building owner, the owner's project manager or technical representative, and/or the facility manager. It is any person who is authorized to make decisions regarding the commissioning project and regularly communicates with the other project team members on the owner's behalf.

Commissioning Authority (CxA) generally refers to an independent, third party commissioning provider. In some instances, it is cost effective and appropriate to assign the duties of the CxA to the qualified facility Management and O&M staff. If this is your case, the CxA will indicate the senior member(s) of your facility Management/O&M staff that is assigned the duties of the CxA. Recommendations on hiring a CxA consultant versus bringing CxA duties in-house are provided for each type of commissioning.

COMMISSIONING FOR NEW CONSTRUCTION AND RENOVATION

Commissioning is a systematic process of ensuring that all building systems and equipment installed as part of new construction or renovation perform interactively according to the design intent and the owner's operational needs. This is achieved

"The pessimist sees difficulty in every opportunity. The optimist sees the opportunity in every difficulty."

Winston Churchill

65

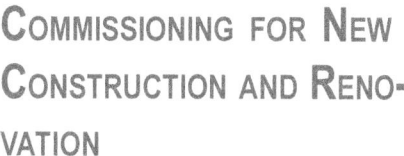

This chapter describes the process for implementing commissioning in new construction and major renovation projects. Chapter 6 covers the retrocommissioning process, Chapter 7 covers the recommissioning process, and Chapter 8 covers continuous commissioning.

In this Chapter

◆ Commissioning for New Construction and Renovation
◆ Pre-Design Phase
◆ Design Phase
◆ Installation / Construction Phase
◆ Acceptance Phase
◆ Post-Acceptance Phase
◆ Best Practices

by beginning in the design phase of a construction project with the documentation of design intent, and continuing through construction, acceptance, and the warranty period with actual verification of performance. The commissioning process encompasses and coordinates the traditionally separate functions of system documentation, equipment startup, control system calibration, testing and balancing, performance testing, and training.

Commissioning typically follows the phases of the new construction or renovation project. Although it is not necessary to perform commissioning tasks during each phase of construction, implementing the process throughout the life of the project will produce the best results. These phases, described in detail in this section, are 1) Pre-design, 2) Design, 3) Installation / Construction, 4) Acceptance, and 5) Post-acceptance / Warranty.

PRE-DESIGN PHASE

Documentation Requirement:
◆ Owner's Criteria

Duties related to commissioning are performed by the owner during the project pre-design phase. The owner hires the architect/engineer (A/E) design team and works with the A/E to determine the design intent and project objectives.

The Owner also determines the commissioning requirements for the project (Will commissioning be part of all phases of the construction project? Will the CxA be brought in only during installation?). The Owner then hires the CxA, preferably through a competitive process.

The Owner's criteria is the only commissioning-related documentation that is developed during pre-design. This document captures requirements of the project, including design objectives and constraints; space, capacity, and performance requirements; flexibility and expandability requirements; and budgetary limitations. The Owner's criteria is used by the A/E to develop the formal design intent and basis of design documentation, and is also used by the CxA to review the design documentation.

The starting point for developing the design intent and basis of design is the identification of the Owner's needs. The design team needs to accurately interpret and record the Owner's vision of the facility itself, how the facility will be used and operated, and the Owner's performance goals and objectives. This vision must be realistic and consider budget restraints, schedules and other limitations.

Technically, considerations should include the use of the facility, the user needs, occupancy requirements, the type of construction, system functions, the expected performance criteria (e.g., energy, air quality, power quality, security, and biohazard and environment), maintainability, supportability, reliability and simplicity.

The Owner will expect the A/E to deliver a design that meets the Owner's identified requirements. The owner will expect the contractor to build the facility in a workmanlike manner in accordance with general accepted construction practices, to use quality materials that are defect-free, to deliver the work on schedule at the agreed price, and to pay the contractor's subcontractors and suppliers in a timely manner (no liens).

DESIGN PHASE

Documentation Requirements:
◆ Project Objectives Document
◆ Design Documentation – Design Intent, Basis of Design, Drawings

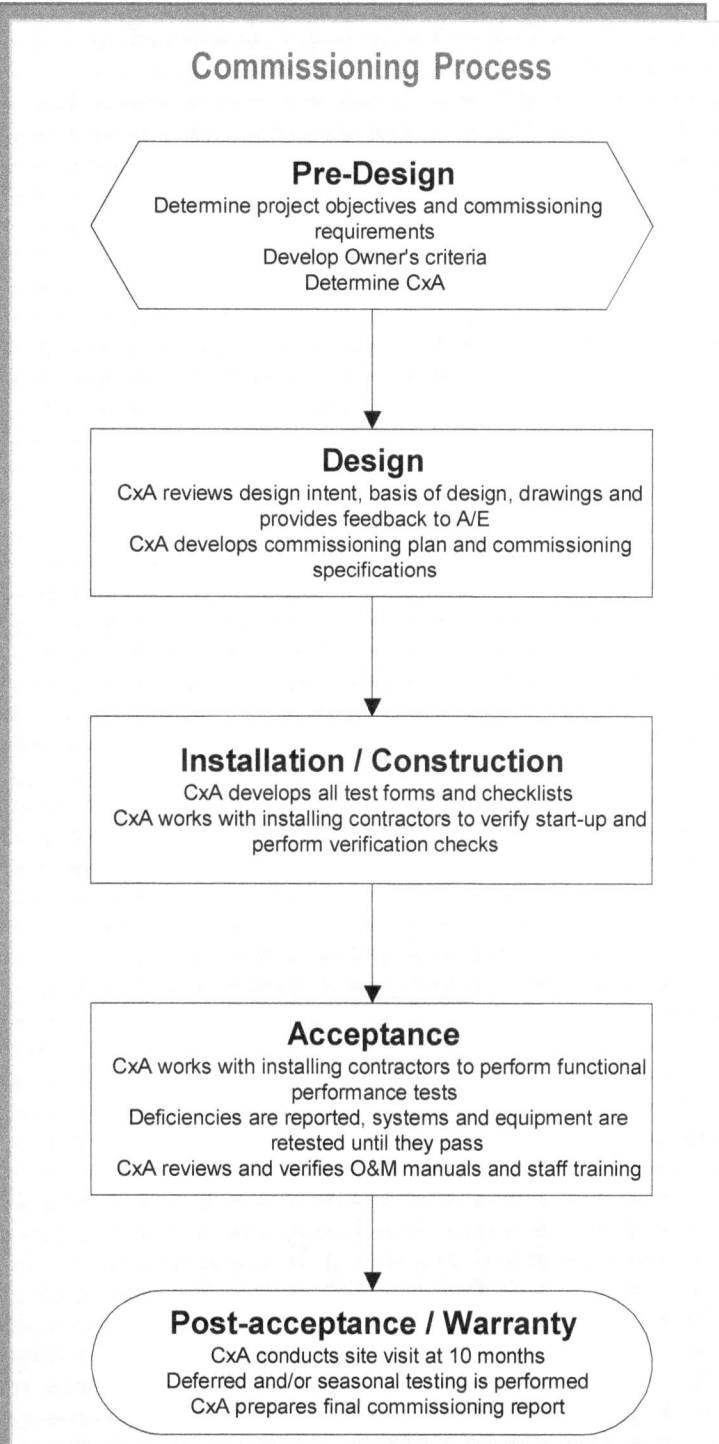

Commissioning Process

Pre-Design
Determine project objectives and commissioning requirements
Develop Owner's criteria
Determine CxA

Design
CxA reviews design intent, basis of design, drawings and provides feedback to A/E
CxA develops commissioning plan and commissioning specifications

Installation / Construction
CxA develops all test forms and checklists
CxA works with installing contractors to verify start-up and perform verification checks

Acceptance
CxA works with installing contractors to perform functional performance tests
Deficiencies are reported, systems and equipment are retested until they pass
CxA reviews and verifies O&M manuals and staff training

Post-acceptance / Warranty
CxA conducts site visit at 10 months
Deferred and/or seasonal testing is performed
CxA prepares final commissioning report

◆ Project and Commissioning Specifications

◆ Commissioning Plan (draft)

During design, the A/E designs the project and produces design documentation, while the CxA reviews the project design for commissioning considerations. The CxA may also coordinate input from the facility users and occupants regarding building features that are of importance to them for integration into the design documentation. The CxA's pres-

Maintainability Factors

1. Accessibility – Consider access to machine room, machine, and machine components.

2. Visibility – Is the component visible from the floor? Nameplate data? Shadow effects?

3. Simplicity – Keep it simple; the more complex, the more difficult and costly.

4. Interchangeability – In an emergency, it is nice to "borrow" a critical part from a less critical component to keep operations up and running.

5. Standardization – Fewer manufacturers and models minimizes spare parts, training, and special tooling.

6. Ease of Monitoring and Testing – Where are the test points located? Gauges? Meters?

7. Human Factors – Can the parts be lifted? Should a chainlift be installed to assist removal of over-sized components, if needed?

8. Safety Concerns – Always paramount.

Supportability Factors

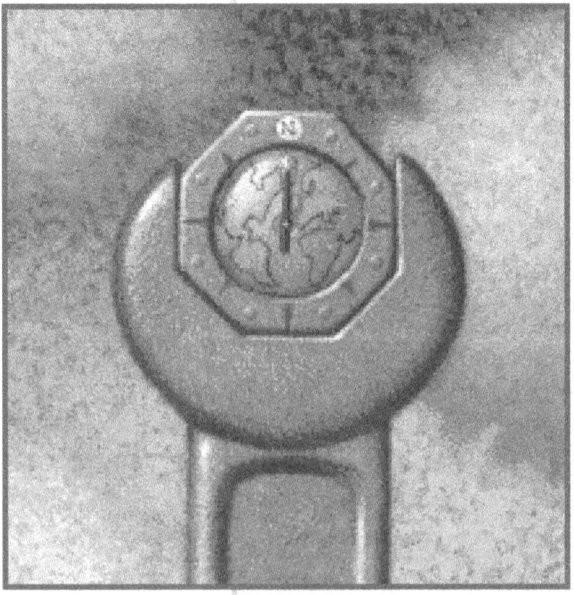

1. Parts Availability – Minimize repair time; consider if parts are locally available or can be ordered via internet for next day delivery. Consider the impact if parts must be special ordered or customized?

2. Repair Capability – Minimize repair time and cost; consider availability of local repair capability. Does a representative need to be flown in from across the country?

3. Training and Skills – Ensure that training and education is available commensurate with the complexity of the equipment being installed.

4. Technical Data – Ensure that all technical support data is available and provided.

ence at bre-bid and pre-construction conferences acquaint potential contractor personnel with the commissioning process.

Project Objectives Document

The project objectives document is prepared by the CxA. This document is based on the Owner's Criteria, and it identifies project requirements relative to energy performance, water performance, LEED certification intentions, maintainability, sustainability, indoor environmental quality, and environmental impacts. This document is used by the CxA as a guide when performing design review for commissioning.

Design Documentation - Design Intent and Basis of Design

The design documentation provides a narrative description of the system or issue, as well as clear and useful background information. Design documentation explains how design and operating objectives will be accomplished. Design documentation includes information from the conceptual design phase and from the design and construction processes, necessary to guide the design, verify compliance during construction, and aid building operations.

Identifying and developing the design intent and basis of design provides each party involved, at each respective state, an understanding of the building systems. This allows team members to perform their respective responsibilities regarding the design, construction, and operation of the building.

Design documentation consists of design intent and basis of design, and the detail of both increases as the design process progresses. At the outset of the project, the design documentation required is primarily a narrative of the building system descriptions, objectives, and how the systems will meet those objectives. As the design process progresses, the design documentation includes the basis of design, a specific description of the systems and components, their function, how they relate to other systems, sequences of operation, operating control parameters, and the assumptions made in the design.

The A/E coordinates the integration of design documentation from each contributing designer to develop the full design documentation by the design team. The A/E, CxA, and Owner each review, comment on, and approve the submissions.

The approved full design intent and basis of design is provided to the CxA at the beginning of the construction phase for use as a guide and baseline reference during start-up, verification, and functional performance testing. Since the job of the CxA is to assure that components and systems have been installed and operate as intended per the Owner's needs, the CxA needs both the Design Intent, based on the A/E's knowledge of the Owner's

69

needs, and the design itself that shows the specified solution. A final as-built copy of the full design intent and basis of design should be prepared and included in the O&M manuals at the end of the construction phase.

Design intent documentation for other systems and components, such as structural, interior design, landscaping, furnishings, etc. may also be required, although not a part of the commissioning process. This should be decided on a project-by-project basis.

Design Intent

The design intent provides an explanation of the ideas, concepts, and criteria that are considered to be important to the owner. It is initially the outcome of the conceptual design phase. The design intent narrative should cover the following, for each system, major component, facility, and area:

1. General system description.
2. Objectives and functional use of the system, equipment, or facility.
3. General quality of materials and construction.
4. Occupant requirements.
5. Indoor environmental quality (space temperature, relative humidity, indoor air quality, noise level, illumination level, etc.).
6. Performance criteria (general efficiency, energy and tolerances of the indoor environmental quality objectives, etc.).
7. Budget considerations and limitations.

8. Restrictions and limitations of system or facility.
9. Special considerations, such as pursuit of LEED-(level) certification.

The design intent should also include design criteria, including at a minimum:

1. Indoor dry bulb temperature and relative humidity.
2. Outdoor dry bulb and wet bulb temperatures.
3. Occupancy, hours of operation, and degree of activity.
4. Lighting and miscellaneous power.
5. Ventilation - recirculation and outside air.
6. Internal and special loads.
7. Insulating R-values for roof, wall, glass, etc.
8. Percentage of glass - fenestration, and types of glass, including coatings and solar coefficients.
9. Building pressurization and infiltration.
10. Building mass.
11. Code requirements and impact on criteria.
12. Air quality design criteria, i.e., ASHRAE 62-91.
13. Noise criteria.
14. Fire and life safety.
15. Energy efficiency and cost.
16. Maintainability.

Basis of Design

The basis of design provides the primary ideas and assumptions behind design decisions that were made to meet the design intent. The basis of design describes the systems, components, conditions, and methods chosen to meet the design intent. It

70

explains how certain systems and space arrangements were chosen by the design team to meet the needs of the Owner.

The following should be included in the basis of design for major equipment:

1. Specific description of systems, components, and methods for achieving the design intent objectives (for instance, why a given system was chosen; details of that system's size, efficiencies, area served, capacity control; integration with other system, sequences of operation under all modes of operation, control strategies, etc.).
2. Equipment maintainability.
3. Fire, life, and safety criteria, strategy narrative, and detailed sequences.
4. Emergency power control and function.
5. Energy performance.
6. Ventilation strategies and methods.
7. Complete sequences of operation, including set points and control parameters.
8. Schedules.
9. Applicable codes and standards.
10. Primary load and design assumptions, including sizing; occupant density and function; indoor conditions such as space temperature, relative humidity,

lighting power density, ventilation, and infiltration rates, etc.; outdoor conditions; and glazing fraction, U-value, and shading coefficient.

The basis of design should also include operations descriptions, including at a minimum:

1. Design intent.
2. Basic system type and major components.
3. Interrelation of components.
4. Capacity and sizing criteria.
5. Redundancy and diversity.
6. Automatic temperature control and sequence of operation.
7. Intended operation under all seasonal loads.
8. Changeover procedures.
9. Part-load strategies.
10. Occupied/unoccupied modes of operation.

11. Design setpoints of control systems with permissible adjustments.
12. Operation of system components in life-safety modes.
13. Energy conservation procedures.
14. Any other engineered operational mode of the system.

Design Documentation Format

The format of the design intent/basis of design documentation should adhere to the following general outline:

1. General design narrative describing the system and/or components.
2. Objectives and functional use of each system and/or components.
3. Full sequence of operations under all modes and conditions.
4. Set points and operating parameters.
5. Performance criteria and applicable codes and standards.

Design Review

The CxA performs reviews of the design intent, basis of design, schematic design drawings and specifications, and the construction document design drawings and specifications, as described in this section.

Design Review Scope

The CxA is not responsible for nor encouraged to check the design for engineering approach, system selection, equipment specification, life cycle costs, or other parts of the overall engineering design that may be construed as second guessing the design engineer. The goal of the CxA's review is to assure that the system can be verified as working correctly and that the system can be maintained in that condition. The CxA reviews the schematic design and construction documents for the following issues at the phases checked for each system commissioned:

- Commissioning Facilitation – Review for effects of specified systems and layout toward facilitating the commissioning process (equipment accessibility for testing, controls, etc.; see *Commissioning Facilitation Review* on page 74).

- Energy Efficiency – Review efficiency of system types and components for specified systems.

- Control System and Strategies – Review specified systems' sequences of operation for adequacy and efficiency.

- Operations and Maintenance (O&M) – Review for effects of specified systems and layout toward facilitating O&M (equipment accessibility, system control, etc.).

- Indoor Environmental Quality – Review to ensure that systems relating to thermal comfort, visual, acoustical, air quality, and air distribution maximize comfort and are in accordance with the

Indoor Air Quality Commissioning Checklist

Design phase IAQ review to be performed:

1. Determine indoor air quality (IAQ) requirements and confirm that these are included in the Project Objectives document.

2. Review expected occupant activity, density, and locations where special attention is needed; review exhaust systems or increased supply air capacity that may be required:

   ❏ Kitchens   ❏ Material storage rooms

   ❏ Break rooms   ❏ Conference rooms

   ❏ Photocopying and/or printing rooms   ❏ Locker rooms

   ❏ Janitorial rooms   ❏ Parking garage

   ❏ Laboratories   ❏ Other: _____

3. Ensure that IAQ objectives are included in the design.

4. Review carbon dioxide sensor locations and control sequences to ensure the system is properly designed to maintain acceptable CO_2 levels.

5. Determine how adequate ventilation rates will be maintained during all occupied modes of operations, particularly during VAV terminal box turn-down.

6. Review air intakes and exhausts for short-circuiting and exterior pollution sources (such as garages, loading docks, and cooling towers).

7. Review potential impact of office partition configurations on ventilation effectiveness.

8. Review filtration type and design, materials, and location.

9. Review HVAC material specifications and application for potential for airflow erosion, corrosion, and microbial contamination (HVAC insulation materials, etc.).

10. Review air supply system to ensure control and minimization of free water and microbial contamination (condensate trays, humidifiers, etc.).

11. Verify the suitability of access doors and inspection ports to all chambers and components of air handling system plenums (accessibility for proper cleaning of both sides of coils, condensate pans and/or humidifier reservoirs, and future duct cleaning).

12. Identify products specified in the contract documents that may contribute to indoor pollutants.

13. Confirm that the specifications specify proper methods and conditions for operating the HVAC system prior to full control and occupancy to minimize dirt and unwanted moisture entering the duct work, coils, building cavities, and any occupied portions of the building.

Project Objectives document and Owner's Criteria. (See *Indoor Air Quality Commissioning Checklist*.)

■ O&M Documentation – Verify adequate building O&M documentation requirements.

■ Training – Verify adequate operator training requirements.

■ Commissioning Specifications – Verify that bid documents adequately specify building commissioning, including testing requirements by equipment type.

■ Project Objectives Document – Verify that the design complies with the project objectives document.

■ Mechanical Systems – Review mechanical design and concepts for enhancements.

■ Electrical Systems – Review electrical design and concepts for enhancements.

Commissioning Facilitation Review

The CxA reviews the design documents for the following issues:

1. Clear design documentation, including detailed and complete sequences of operation for specified equipment.
2. HVAC fire response matrix that lists all equipment and components (air handling units, dampers, valves, etc.) with their status and action during a fire alarm or emergency.
3. Maintainability and supportability.
4. Required isolation valves, dampers, interlocks, piping, etc. to allow for manual overrides, simulating failures, seasons, and other testing conditions.
5. Sufficient monitoring points in the building automation system (BAS), including those beyond that necessary to control the

74

Sample Organization of Commissioning Specification

1.1 Commissioning Description

1.2 Commissioning Project Coordination

1.3 Commissioning Process

1.4 Related Work

1.5 Responsibilities

1.6 Definitions

1.7 Systems to be Commissioned

2.1 Test Equipment

3.1 Meetings

3.2 Reporting

3.3 Submittals

3.4 Startup, Verification, and Initial Checkout

3.5 Phased Commissioning

3.6 Functional Performance Testing

3.7 Documentation, Non-Conformance, and Approval of Tests

3.8 Operation and Maintenance (O&M) Manuals

3.9 Training of Owner Personnel

3.10 Deferred Testing

3.11 Written Work Products

systems, to facilitate performance verification and O&M.

6. Adequate trending and reporting features in the BAS.

7. Pressure and temperature (P/T) plugs close to controlling sensors for verifying calibration.

8. Pressure gauges, thermometers, and flow meters in strategic areas to facilitate verifying system performance and ongoing O&M.

9. Adequacy and location of vibration sound discs/test points and lube oil sampling ports, as applicable.

10. Rationale for location for the variable air volume (VAV) duct static pressure sensor and chilled water differential pressure sensor.

11. Adequate balancing valves and flow meters to facilitate reliable testing and balancing (TAB) of the HVAC systems.

12. Uniform inlet connection requirements to VAV terminal boxes.

13. Energy efficiency and reduced emissions in support of current Federal energy statutes and Executive Orders.

14. Maximization of points toward LEED certification, as applicable.

15. Clear and complete commissioning specifications for the construction phase.

16. Complete O&M documentation requirements in the specifications.

17. Complete training requirements in the specifications.

Commissioning Specification

The CxA prepares commissioning specifications specific to the project requirements in Divisions 1, 15, 16, and 17. Commissioning specifications that should be prepared and added to the project specifications include:

- SECTION 01810 – Extra General Requirements.
- SECTION 15995 – Mechanical Systems Commissioning (including plumbing and HVAC).
- SECTION 16995 – Electrical Systems Commissioning (including fire alarm and protection).
- SECTION 17100 – Commissioning Requirements.

The commissioning specifications should be reviewed and commented on by the Owner. Each should include, as a minimum, the following:

1. Detailed description of the responsibilities of all parties.

2. Detailed description of the commissioning process.

3. Reporting and documentation requirements, including formats.

4. Alerts to coordination issues.

5. Deficiency procedures and resolutions.

6. Construction checklists and start-up requirements.

7. Functional performance testing procedures, organized by equipment and system.

8. Specific functional performance testing requirements, including testing conditions and accep-

tance criteria for each piece of equipment being commissioned.

Commissioning Plan

The CxA develops the commissioning plan during design, but the plan is updated and revised as necessary throughout the construction phase to reflect any approved changes or equipment substitutions. The commissioning plan provides additional guidance to project team members in the execution of the commissioning program.

The following information, as a minimum, should be included in the commissioning plan:

Overview and General Information

- General Building Information
- Definitions and Abbreviations
- Purpose of the Commissioning Plan
- Commissioning Scope
- Equipment and Systems to be Commissioned

Construction and Commissioning Team Members

- Points of Contact

- Project Organization Chart

Roles and Responsibilities

- General Management Plan
- Roles and Authority
  - All Parties
  - Commissioning Authority
  - Owner / Project Manager
  - Architect / Engineer Design Team
  - General Contractor / Construction Manager
  - Building O&M Personnel
  - Manufacturers and Vendors

Commissioning Plan

- Commissioning Scoping Meeting
- Scheduled Commissioning Meetings
- Management and Communication Protocols
- Progress Reporting and Logs
- Site Observation
- Initial Submittals and Documentation
  - Standard Submittals
  - Special Submittals, Notifications and Clarifications
- Development of Functional Test and Verification Procedures
  - Scope of Testing
  - Development Process
- Verification Checks, Tests and Startup
  - Execution
  - Sampling Strategy
  - Deficiencies and Non-Conformance
  - Testing, Adjusting, and Balancing (TAB)
  - Controls Checkout Plan
- Functional Performance Testing
  - Overview and Process
  - Sampling Strategy
  - Deficiencies and Retesting

- Facility O&M Staff Participation
- O&M Manuals and Warranties
  - Post of System Operating Instructions
  - Standard O&M Manuals
  - Commissioning Record
- O&M Orientation and Training
  - Training Requirements
  - Schedule
- Warranty Period

Written Work Products

- List of deliverables

Commissioning Schedule

- Project and commissioning schedules, updated regularly

Appendices

- Design Intent and Basis of Design
- Verification Checklists
- Functional Performance Test Forms

Pre-Bid Meeting

Often bidders will have questions regarding their roles in the commissioning process. These questions should be answered by the CxA at the pre-bid meeting. Contractors generally accept the process much more readily if they understand it and if the CxA exhibits a positive, helpful, and cooperative approach right from the start.

INSTALLATION / CONSTRUCTION PHASE

Documentation Requirements:
- Verification Checklists
- Functional Performance Test Forms
- Commissioning Progress Reports / Deficiency Logs
- Controls Checkout and TAB Plans
- Commissioning Plan (update as necessary)

During construction, the installing contractors (mechanical, HVAC, electrical, controls, etc.) install the designed equipment and systems under the direction of the owner and the owner's general contractor or construction manager (GC/CM). The CxA develops commissioning test procedures and forms, witnesses equipment and systems start-up, and verifies that equipment and systems are ready for functional performance testing performed during the acceptance phase. Deficiencies are tracked and reported, and the CxA schedules re-testing until the equipment and systems are ready for functional performance testing.

Commissioning Meetings

The CxA keeps minutes from all commissioning meetings for distribution to the commissioning team. It is the CxA's responsibility to schedule and conduct commissioning meetings.

Commissioning Scoping Meeting

The CxA conducts a commissioning scoping meeting soon after the start of construction. In attendance are the identified commissioning team members and other key individuals from the project team (Owner,

CxA, GC/CM, A/E design team, installing contractors, TAB, facility O&M, manufacturer/vendor representatives). The commissioning plan, process, and schedule are reviewed by the CxA, including each party's role and responsibilities.

The outcome of the meeting is an increased understanding by all parties of the commissioning process and their respective responsibilities. The meeting provides the CxA additional information needed to finalize the commissioning schedule.

Commissioning Meetings

The CxA holds regular commissioning team meetings during the project construction and acceptance phases. Regular commissioning meetings are intended to accomplish the following:

■ Review and update commissioning schedule based on any changes to the project schedule.
■ Report observed deficiencies or other problems, discuss problem

resolution.
■ Review with commissioning team members commissioning responsibilities and answer any questions or concerns they may have on the commissioning process.

The frequency of the commissioning meetings depends on the project schedule, but meetings should be held at least bi-weekly during start-up and functional performance testing.

Controls Coordination Meeting

The CxA may schedule and conduct a meeting to address integration issues between equipment, systems, and disciplines to ensure that integration issues are addressed and responsibilities are clearly defined. Integration issues and suggested remedies/responsible parties should be documented and submitted to the Owner and GC/CM.

Project Meetings

The CxA also attends regular project progress meetings, if possible, in order to remain informed on the project progress and to update parties involved in the commissioning. Project meetings provide the CxA with information on substitutions, change orders, and Architect's Supplemental Instructions (ASI) that may affect commissioning equipment and systems or the commissioning schedule. (It is the responsibility of the GC/CM to provide this information to the CxA if the CxA misses a project meeting.)

The CxA reviews construction meeting minutes, change orders, and Requests for Information (RFI) for the same purpose.

Submittals Review

As the contractors make equipment submittals to the design team and project manager, copies are routed to the CxA for information. The CxA is not responsible for approving the submittals (that is the designer's responsibility), but the CxA reviews submittals for applicable systems being commissioned for compliance with commissioning requirements. The CxA reports to the design team if there is anything in the submittals that appears seriously wrong.

Verification Checklists

Verification checklists are developed by the CxA to guide the start-up process and ensure that equipment and systems are prepared for functional performance testing.

Verification checklists are primarily static procedures to prepare the equipment or system for initial operation. In general this includes factors such as:

- Equipment is located according to plans and practicality, such as full accessibility for maintenance.
- Lubrication levels are acceptable.
- Fan belt tension is in accordance with manufacturer requirements.
- Labels are affixed.
- Gauges are ergonomically in place.
- Correct valves are installed and are accessible.
- Sensors and instrumentation are calibrated and installed per the engineering requirements (such as duct pressure sensors located three-fourths of the distance to the furthest point in a VAV duct system).
- Correct interlocks and interfacing between HVAC equipment, systems, subsystems, and other building systems.
- Drain piping is properly sloped.
- Proper sheave alignment, connection to power and other utilities, vibration isolation, and pipe and duct support.
- Completion of testing and balancing (TAB) work.

Some verification checklist items entail simple testing of the function of a component, a piece of equipment, or system (such as measuring the voltage imbalance on a three-phase pump motor of a chiller system). Verification checklists augment and are combined with the manufacturer's start-up checklist.

Verification checklists are important to ensure that the equipment and systems are hooked up and operational and that functional performance testing may proceed without unnecessary delays. In general, the verification testing for a given system must be successfully completed prior to formal functional performance testing of equipment or subsystems of the given system.

A sample Verification Checklist is provided in Appendix A, Sample Commissioning Forms.

Many of the problems found during verication are associated with equipment maintainability and accessibility. Most common are:

■ Equipment cabinet doors blocked by piping or structural members.
■ Ceiling spaces are too crowded to allow access to the equipment.
■ Terminal air distribution devices installed too high above suspended grids to allow safe access by ladder.

■ Balance valves not installed.
■ Pump flow fittings and gauges located too close to pipe bends, suction diffusers, or other pipe characteristic that makes their reading inaccurate.
■ Equipment nameplate data is not visible.

Installing contractors typically already perform some, if not most, of the verification checklist items the CxA will recommend. However, few contractors document in writing the execution of these checklist items.

Development of Verification Checklists

The CxA requests and reviews relevant information prior to system start-up and verification, including O&M materials and manufacturer's start-up and check-out procedures.

Before start-up, the CxA gathers and reviews current control sequences and interlocks and works with each installing contractor and the design team to verify that the functional testing procedures that are in the commissioning specifications are appropriate.

The CxA writes and distributes verification checklists to each installing contractor for equipment to be commissioned. The original checklists are often organized in a binder and left at the job site in the custody of the GC/CM so that they are accessible and can be annotated as work completion progresses.

Case Study: Commissioning in Action

An A/E firm formed a joint venture to design an energy plant to serve two hospitals. The facility was to produce and pump chilled water through a network of pipes and coils within the air-conditioning system of both hospitals. The plant experienced several pump failures, pipe fractures, and excessive energy consumption by the air-conditioning system. This led to a claim by the hospitals against the joint venture. To mitigate the damages, the joint venture agreed to commission the plant. After an extensive review, the joint venture was able to prove that the alleged damages were the result of ineffective maintenance, defective maintenance systems, and failed standard service items. The claim was settled for $30,000.

Had commissioning services been provided during construction, the team probably would have identified potential system failures prior to occupancy. Neither the Owner nor the joint venture would have had claim-related expenses. The commissioning would have provided a documented benchmark against which future system performance could be trended and compared. (Source: *Planning to Avoid Commissioning and Facility Management Claims*; Constructive Comments, Number 4, 2005, Victor O. Schinnerer & Company, Inc., www.Schinnerer.com)

Personal digital assistants (PDAs) offer an alternative to paper checklists. The checklists are produced electronically and downloaded onto the PDA. Data is entered onto the PDA directly, real time as the testing progresses, then downloaded in a permanent file for archiving and analysis. This eliminates the need to maintain paper or reenter data redundantly into the computer, which increases the likelihood of human error.

Execution of Verification Checklists

Four weeks prior to start-up, installing contractors and vendors schedule start-up and initial checkout with the GC/CM and CxA. The start-up and initial checkout are directed and executed by the installing contractor or vendor using the checklists provided by the CxA. The CxA observes and validates the results for each type of primary equipment.

To document the process of startup and checkout, the site technician performing the line item task checks off items on the verification and manufacturer field checkout sheets as they are completed. The installing contractors and vendors execute the checklists and submit a signed copy of the completed verification checklists to the CxA.

On smaller equipment or projects, the checklists (which contain more than one trade's responsibility) may be passed around to the contractors to complete. For larger equipment, each trade may need a full form

and the CxA consolidates the forms later.

The CxA documents systems start-up by reviewing each installing contractor's completed verification checklists and by selected site observation. Site visits are conducted by the CxA during equipment installation to verify the commissioned equipment and systems are installed according to the manufacturer's recommendations and to industry accepted standards, and that equipment has received adequate operational checkout by the installing contractors. The CxA also witnesses a sampling of ductwork testing and cleaning to be confident that proper procedures have been followed.

Documentation of all system start-up and verification activities should be included in the Commissioning Report (described in the *Post-Acceptance / Warranty* section).

Deficiencies and Non-Conformance

The installing contractors clearly list at the bottom of the procedure form or on an attached sheet any outstanding items of the initial start-up and verification procedures that were not completed successfully. The respective contractor provides the procedure forms and deficiencies to the CxA within two days of test completion. The CxA works with the installing contractors and vendors to correct and retest deficiencies or uncompleted items, involving the GC/CM, if necessary.

81

The installing contractors or vendors correct all areas that are deficient or incomplete according to the checklists and tests. The CxA recommends approval of the start-up and initial checkout of each system to the Owner and GC/CM.

Commissioning Progress Reports / Deficiency Logs

The CxA provides the Owner and GC/CM with weekly commissioning progress reports that include:

- An update of the commissioning schedule, including schedule changes and new items added to the schedule.
- An update of the commissioning progress (start-up verified and complete).
- A list of new and outstanding deficiencies.
- A list of deficiencies that have been resolved.

Controls Checkout Plan

The controls contractor develops and submits a written step-by-step plan to the CxA that describes the process to be followed in checking the control system and the forms to be used to document the process.

The controls contractor also meets with the testing, adjusting, and balancing (TAB) contractor prior to the start of TAB. They review the TAB plan to determine the capabilities of the control system for use in TAB. The controls contractor provides the TAB contractor with any necessary unique instruments for setting terminal unit boxes (i.e., hand held control system interface for use around the building during TAB) and instructs the TAB contractor in their use.

The controls contractor also provides a technician qualified to operate the controls to assist the TAB contractor in performing TAB.

Prior to substantial completion, the CxA reviews detailed software documentation prepared by the controls contractor for all direct digital control (DDC) systems. This includes reviews of vendor documentation, the programming approach, and the specific software routines applied to project facility components and building systems.

All required controls verification checklists, calibrations, and start-up of the system should be completed and approved by the CxA prior to TAB. The controls contractor ex-

ecutes the assigned tests and trend logs, and remains on site for assistance for mechanical system functional tests as specified. The CxA verifies and documents the effective operation of these interlocking control systems.

Testing, Adjusting, and Balancing (TAB) Plan

The TAB contractor submits the outline of the TAB plan and approach to the CxA and the controls contractor eight weeks prior to starting the TAB. Included in the approach is an explanation of the intended use of the building control system during testing. The CxA reviews the plan and approach for understanding and coordination issues and may comment The controls contractor reviews the feasibility of using the building control system for assistance in the TAB work.

The TAB contractor submits weekly written reports of discrepancies, contract interpretation requests, and lists of completed tests to the CxA and GC/CM. This facilitates quicker resolution of problems and will result in a more complete TAB before functional testing begins.

Commissioning Plan

The CxA continues to update and revise the commissioning plan throughout the construction phase to reflect any approved changes or equipment substitutions.

Functional Performance Test Forms

The CxA oversees functional performance testing during the acceptance phase, but the test plans and forms are developed during the construction phase (or earlier). Functional testing includes operating the system and components through the significant modes of operation, including:

1. Each of the written sequences of operation.
2. Start-up and shut-down.
3. Unoccupied mode.
4. Manual mode.
5. Staging.
6. Miscellaneous alarms.
7. Power failure.
8. Interlocks with other systems or equipment.

The systems performance is evaluated for:

- Input and output capacities.
- Flow and distribution performance.
- Control system performance, accuracy, and adherence to sequences of operation.
- Minimum or part/load operations and capabilities.
- Interface with other equipment and/or systems.
- Emergency response

Actual physical responses must be observed. Reliance on control signals and other indicators is unacceptable.

A sample Functional Performance Test form is provided in Appendix A, Sample Commissioning Forms.

84

Equipment and system functional performance testing begins only after the affected systems have been fully and successfully verification tested and the system has been tested and balanced (TAB). Sensors and actuators are calibrated by the installing contractors prior to functional testing, and checked by the CxA. Functional testing is performed using conventional manual methods, control system trend logs, and stand-along data loggers to provide a high level of confidence in proper system function.

Systems and subsystems are tested under full load, where possible and applicable, and under part load conditions as specified in the Cx Plan.

Tests on HVAC equipment are performed during both heating and cooling seasons (see *Deferred and Seasonal Testing*).

Development of Test Forms and Procedures

The CxA, based on input from other team members, develops the functional performance test procedures. This is to ensure all aspects of system operation are fully explored and documented. The CxA obtains clarification, as needed, from installing contractors and the A/E regarding the sequences of operation. Prior to execution, the CxA provides a copy of the primary equipment tests to each installing contractor or vendor, who reviews the tests for feasibility, safety, warranty, and equipment protection. Blank copies of the procedures are included in the O&M

manuals for later use by operations staff.

The CxA reviews the factory or required Owner acceptance tests and determines what further testing may be required to comply with the specifications.

The CxA reviews proposed testing procedures and report formats and observes sufficient field testing to confirm that all I/O points have been properly tested.

ACCEPTANCE PHASE

Documentation Requirements:
◆ Commissioning Progress Reports / Deficiency Logs
◆ O&M Manuals
◆ Commissioning Plan (finalized)
◆ Commissioning Record

During acceptance, the installing contractors perform functional performance tests on equipment and systems, which are witnessed by the CxA. Deficiencies are tracked and reported, and the CxA schedules retesting until the equipment or systems operate and interact as designed. Operations and maintenance (O&M) manuals are organized and reviewed, and O&M staff training is scheduled and executed. The project is finalized and handed over to the owner.

Functional Performance Testing

The CxA schedules, oversees, witnesses, and documents the functional performance and condition

---

Test Sampling and Deficiencies

Multiple pieces of identical equipment can be functionally tested using a sampling strategy. In functional performance testing, should 10% of the sampled systems verified fail to meet the design criteria, another 10% should then be tested. If those additional 10% of the systems also fail to meet the design criteria, the entire system in question should be re-examined by the responsible contractor. The cost to commission that particular system should be borne by the responsible contractor.

---

acceptance testing of all equipment and systems according to the commissioning specifications and the Commissioning Plan. The installing contractors or vendors execute the tests.

The control system is tested before it is used to verify performance of other components or systems. The air balancing and water balancing is completed and debugged before functional testing of air- or water-related equipment or systems. Testing proceeds from components to subsystems to systems and finally to interlocks and connections between systems. In addition to the verification of sequences during normal operating conditions, testing also includes any abnormal scenarios during which the equipment may be expected to operate, such as during failure and recovery, standby power, and alarm and alert situations.

The CxA also observes, photographs or video captures, and documents the actual performance of safety shutoffs in real or closely simulated failure condition.

A sample of redundant items included in the contractor's test and balance (TAB) report is checked

for accuracy as prescribed in the Cx Plan. If a substantial failure rate is encountered, all should be corrected and a larger and different sample chosen for a repeat test at the contractor's expense. For example, 20 percent of all terminal devices such as grilles and registers may be selected for output verification. If the output of 10 percent of these differ significantly from the reported values, all should be rejected, corrective action taken by the contractor, and a new sample of 25 percent (or more) randomly selected devices selected for re-verification.

During the functional performance testing, the CxA looks for four features relative to the test data:

- **Repeatability** – How well does the component consistently replicate a desired output, such as control sequences and resulting pressure or flow?

- **Stability** – How well does the component or system maintain a desired condition despite changing outside influences?

- **Responsiveness** – How well do the components work

together in an integrated fashion to react to changing outside influences while maintaining the desired outcome?

■ **Accuracy** – Does the system achieve the desired outcome within an acceptable tolerance?

When the functional performance testing is completed, every mode of each operation of a system, each piece of equipment, every item in the control sequence description, every emergency response, and every zone or subsystem will have been proven to operate as required and specified in the design intent and design documents.

Deficiencies and Re-testing

The CxA documents the results of each functional performance and condition acceptance test. Corrections of minor deficiencies (e.g., fixing a controller, adjusting alignment) identified can be made by the respective contractor during the tests at the discretion of the CxA.

In general, no applicable systems or subsystems are accepted until all items of equipment have been successfully functionally performance and condition tested. After all deficiencies have been corrected, the entire functional performance test for the equipment, system, or subsystem is repeated.

The CxA records the results of the test on the procedure or test form. Deficiencies or non-conformance issues are noted and reported to the Owner and GC/CM. Installing contractors correct the noted deficiencies and notify the CxA, who schedules re-testing.

Decisions regarding deficiencies and corrections are made at as low a level as possible, preferably between CxA, GC/CM, and installing contractors. For areas in dispute, final authority, besides the Owner's, resides with the A/E design team. The CxA makes final recommendations to the Owner for acceptance of each test after a review of the final functional performance test. The Owner gives final approval on each test.

Building O&M Staff Participation

The building's O&M staff is strongly encouraged to attend and participate in the testing process.

Condition Acceptance Testing

Some CxAs integrate condition acceptance testing into their commissioning programs as part of the functional performance testing. Commissioning uses verification and functional performance testing to ensure that systems are installed correctly and operate properly, and to validate output parameters such as air output, flow, temperature, pressure, system balance, control sequences, and energy consumption. The entire system is evaluated through all control modes and includes testing of the ductwork, piping, valves, and actuators.

86

Commissioning does not evaluate, however, the actual *condition* of the equipment. Whereas functional performance testing will normally tell you that a fan is pushing out a certain CFM, it will not tell you if the fan and driver are properly aligned, if the fan is balanced, or if there are loose, high resistance electrical connections. This is where a reliability centered maintenance-based condition acceptance element to the commissioning program comes in.

Condition acceptance testing may be integrated with the functional performance testing. The results are that:

- Latent manufacturing and installation defects are detected early in the process while the contractor is still on site.
- Warranties are enforceable by documenting the precise condition and performance of the system at the time of acceptance.
- Because problems are detected and corrected at the system, rather than component level, synergy is assured, resulting in greater system reliability, fewer recurring problems after acceptance, and reduced life cycle costs.
- The baseline documentation generated becomes the foundation of the maintenance program for the life of the facility.

Condition acceptance testing uses technology such as vibration monitoring, infrared thermography, insu-lation resistance testing, motor current analysis, ultrasound, and oil analysis (as appropriate). It evaluates the installed system to ensure that the equipment is properly aligned and balanced; that the correct lubricants are provided, are clean, and have the correct required additives; that electrical connections are tight and proper; that motors are free of internal defects and are phase balanced and have insulation resistance within specific tolerances; that the equipment has no internal damage from the factory, from handling and transport, or from the installation; or that the installation meets ISO quality standards.

Rarely are these factors looked at during "equipment start up" and inevitably, they become problematic for the maintenance staff the day after the warranty expires. Condition acceptance testing provides a baseline for each unit of equipment against which future data from condition monitoring can be compared, trended, and an alert made before equipment failure. Necessary repairs, then, can be scheduled at an opportune, non-critical time after parts and materials have been ordered and received and labor has been scheduled.

Commissioning Progress Reports / Deficiency Logs

The CxA continues to provide the Owner and GC/CM with weekly commissioning progress reports that include:

- An update of the commissioning schedule, including requested

87

schedule changes and new items added to the schedule.

- An update of the commissioning progress (functional performance and condition acceptance tests completed).
- A list of new and outstanding deficiencies.
- A list of deficiencies that have been resolved.

Operation and Maintenance Manuals

The CxA compiles and reviews the operation and maintenance data provided by installing contractors, manufacturers, and vendors for thoroughness. The CxA also gathers and reviews as-built drawings for equipment and systems that were commissioned to verify compliance with the specifications. The CxA takes the lead in inspecting and approving the O&M manual as to content and organization. The following information should be included in the final O&M manuals:

1. Project objectives document.
2. Design intent and basis of design documents.
3. Sequences of operation (updated to as-built from contract documents), equipment operation schedules, and point lists with initial set-points and ranges.
4. TAB data.
5. Instruction for operation for each piece of equipment, including seasonal adjustment, start-up and shut-down checklists, and instructions for energy savings operations and strategies.

6. Functional performance test procedures and results, blank test forms, and recommended schedule for re-testing.
7. Recommendations for recalibration frequency for sensors and actuators by type and use.
8. Single-line diagrams of commissioned systems (as-built from construction documents).
9. Troubleshooting table for ongoing achievement of project objectives.
10. Guidelines for continuous maintenance of the project objectives, design intent, and basis of design.
11. Equipment warranties information, including warranty start and end dates, and verification that all requirements to keep the warranty valid are clearly stated. All warranty information is consolidated in a single section of the O&M manuals.
12. A consolidated listing of all manufacturer and vendor points of contact, including representative name, company name and address, phone numbers, e-mail addresses, and web address for quick reference.

It is good practice to aim for the completion of the O&M manual as soon as possible after the last contractor submittal is approved. By this time, all equipment should have been ordered and O&M information accumulated. Each contractor should have assembled and completed the O&M manuals by about the time all equipment is physically on site. Consequently, receipt of the completed O&M manuals should be a factor in approving each contractor's 60- to 90-percent progress payments.

Final Commissioning Plan

The CxA finalizes the Cx Plan after functional performance testing is complete and there are no more changes or equipment substitutions planned.

Commissioning Record

The CxA compiles, organizes, and indexes commissioning data by equipment for inclusion in the O&M manuals. The correspondence, meeting minutes and progress reports, miscellaneous notes, etc. kept during the project will not be retained into this record, but stored in a separate volume.

Operation and Maintenance Staff Training

Operations and maintenance staff training actually begins early in the commissioning process through the participation of select representatives on the commissioning team. As the project progresses, the O&M staff becomes more and more acquainted with the building's features. Designated representatives visit the job site during selected time periods including:

- System installation.
- System verification.
- Hands on equipment startup.
- Hands on functional performance testing.
- Hands on condition acceptance testing.

89

- Formal manufacturer and installer training sessions.

It is highly advantageous for the maintenance technicians to see cabling, piping, ductwork, and other normally hidden systems prior to their being covered and obstructed by insulation, ceilings and drywall. In addition, the O&M staff must be walked through emergency procedures and various operational sequences under all possible scenarios.

The O&M staff must become familiar with the building's control system:

- How the control system works, including the sequence of operations.
- The control system structure – what is connected to what.
- The control system components and their proper nomenclature (e.g., actuators, sensors, valves, terminal boxes).

- The control system monitor, its various screens, and the information the software is capable of providing.
- Alerts, alarms, and emergency procedures.

The CxA is responsible for overseeing and approving the content and adequacy of formal training of O&M personnel for commissioned equipment. The GC/CM is responsible for training coordination and scheduling and ultimately for ensuring that training has been completed. The GC/CM and installing contractors actually present the training material. The CxA attends all training sessions and sees to it that important issues are raised.

Each installing contractor and vendor responsible for training submits a written training plan to the CxA for review and approval prior to training. The plan should cover the following elements:

- Equipment included in training.
- Intended audience.
- Location of training.
- Objectives.
- Subjects covered (description, duration of discussion, special methods, etc.).
- Duration of training on each subject.
- Instructor for each subject and instructor's qualifications.
- Methods (classroom lecture, video, site walk-through, actual operational demonstrations, written handouts, etc.).

The CxA assists the GC/CM and Owner in developing an overall

training plan and in coordinating and scheduling the overall training for the commissioned systems. The CxA develops criteria for determining that the training has been satisfactorily completed, including attending some of the training.

The CxA recommends approval of the training to the Owner, who gives approval when acceptable training has been conducted.

Scheduling a brief presentation by the design engineer is a good practice that informs the O&M staff of the idiosyncrasies and features of the installed systems. It also allows the staff to ask questions that remain after the construction process and were never really completely clear on the design drawings.

Video-taping the training is another good practice for archiving the information and for training new employees as they come on board. Good sound is as important as good picture, so if hands-on and other training is taking place in the actual equipment rooms, then some mitigation of equipment noise (such as turning off air compressors) may be required.

Special Training and Orientation

The following are additional special training and orientation sessions that may be requested by the Owner:

- **Recommissioning** - The CxA provides instruction on the use of blank functional test forms for periodic recommissioning of equipment and systems, per the specification.

- **Architect/Engineer** - The A/E provides a general overview of the facility, its use, special features, tenant and public considerations, etc.

- **Mechanical Design Engineer** - The mechanical designer provides an overview of the major systems and equipment in the facility, including for each system: the design intent, why the system was chosen, an overview of its operation and interactions with other systems, any special areas to be aware of, issues regarding future expansion and remodeling, etc.

- **Electrical Design Engineer** - The electrical designer provides an overview of the major systems and equipment in the facility, including for each system: the design intent, why the systems was chosen, an overview of its operation and interactions with other systems, any special areas to be aware of, issues regarding future expansion and remodeling, etc.

- **Vendors and Manufacturers** - The vendor or manufacturer of any item that is commissioned provides an overview of that particular piece of equipment in the facility, including for each item its physical make-up, capabilities and interactions with other systems, test points for condition monitoring (if applicable), any special areas to be aware of, and issues regarding expansion capabilities or adding capacities.

POST-ACCEPTANCE / WARRANTY PHASE

Documentation Requirement:
◆ Final Commissioning Report

The CxA returns to the project approximately 10 months into a 12-month warranty period. During this visit(s), the CxA reviews with the facility staff the quality and reliability of the current building operation. The status of outstanding issues related to the original and seasonal commissioning is also addressed. The CxA interviews facility staff and to identify problems or concerns that they have operating the building as originally intended. Issues discussed may include:

1. Current building operation.
2. Any outstanding issues related to the project and the commissioning process.
3. Any problems or concerns with operating the building as originally intended and designed.
4. Suggestions for improvements and for recording enhancements in the O&M manuals and final Commissioning Report.
5. Areas that may come under warranty or under the original construction contract.
6. Documentation of reports and documents, and requests for services to remedy outstanding problems.

Deferred and Seasonal Testing

During the warranty period, seasonal testing (tests delayed until

91

weather conditions are closer to the system's design) are scheduled and executed under the coordination of the CxA. Tests are executed, documented, and deficiencies corrected by the appropriate contractors, with building O&M staff and the CxA witnessing. Any final adjustments to the O&M manuals and as-built drawings due to the testing should be made by the responsible contractor.

If any check or test cannot be completed due to the building structure, required occupancy condition, or other deficiency, execution of functional performance testing may be delayed upon approval of the Owner. These tests will be conducted as soon as possible in the same manner as the seasonal tests, and the services of necessary parties will be negotiated.

Final Commissioning Report

The final commissioning report includes a summary report of participants and their roles, building description, project objectives, an overview of the commissioning and testing scope, and a general description of testing and verification methods.

For each piece of commissioned equipment, the report contains the disposition of the CxA regarding the adequacy of the equipment, documentation, and training relative to the contract documents in the following areas:

- Equipment meeting project and commissioning specifications.
- Equipment installation.
- Functional performance and efficiency.
- Equipment documentation and design intent.
- O&M staff training.

All outstanding non-compliance items should be specifically listed. Also included are training records, test schedules, construction checklists, and recommendations for improvement to equipment or operations, future actions, and commissioning process changes. Each non-compliance issue is referenced to the specific functional test, inspection, trend log, and design specification requirement where the deficiency is documented.

The functional performance and efficiency section for each piece of equipment includes a brief description of the verification method used (e.g., manual testing, BAS trend logs, and/or data loggers), and observations and conclusions from the testing.

The Final Commissioning Report is a collection of project documentation. Typically, it includes:

- Design Intent and Basis of Design
- Commissioning Plan
- Signed checklists
- Signed functional performance test results
- Requests for information (RFIs)
- Deficiency reports / Corrective actions
- TAB data

92

- Equipment condition baseline data
- Planned off-season testing

BEST PRACTICES

- In nearly all cases, it is easier and in the long run more cost effective to hire an independent, third-party commissioning authority to guide the commissioning project.

- Ongoing monitoring and verification of energy and operational performance are essential to maintaining persistence of improvements and ensuring that equipment and systems are operating at optimal efficiency.

- A well-trained O&M staff with adequate resources is crucial to the success of any commissioning program.

- Problems, deficiencies, and complaints should be carefully recorded by the facility management and O&M staff. Often, problems can be grouped into categories (design, operation, maintenance, installation, comfort and safety) that can be analyzed for more significant trends.

- The documentation required by each commissioning process can be daunting, but will prove an invaluable resource for building management, O&M, and future commissioning efforts.

93

STUDY QUESTIONS

1. What are the commissioning documentation requirements for each phase of construction? Why are these documentation requirements so important to the overall commissioning program?

2. What is the role of the CxA in verification and functional performance testing? What is the role of the installing contractors? The CM/GC?

3. What are some factors to bear in mind when evaluating maintainability and supportability during design review?

4. How are the design intent and basis of design different? What part does the Owner's project requirements document play in both the design intent and basis of design?

5. What issues should be checked for each commissioned system during design review?

6. What commissioning facilitation issues should be addressed during design review?

7. How is the commissioning specification different from the commissioning plan? Which party(ies) prepares each document?

8. Who conducts commissioning meetings during the installation/construction phase of the project? How do these meetings relate to regular construction team meetings?

9. When is verification testing performed, and by whom? When is functional performance testing performed, and by whom? When is deferred and/or seasonal testing performed, and by whom?

10. What is the process for reporting and correcting deficiencies and non-comformance issues?

11. How does the TAB contractor fit into the commissioning program on a construction project?

12. What are some of the tools and instrumentation used by the installing contractors and CxA to accomplish functional performance testing?

13. When should test sampling be utilized during functional performance testing?

14. What are the benefits of performing condition acceptance testing in addition to functional performance testing?

Chapter 6
Retrocommissioning Process

Retrocommissioning is a systematic process for improving and optimizing building performance in an existing building that has never gone through any type of commissioning or quality assurance process. Its focus is usually on energy-using equipment such as mechanical equipment, lighting, and related controls. However, building envelope is an increasing retrocommissioning issue, particularly in humid areas where mold is a particular problem.

The most common reasons why an Owner may want to retrocommission a facility include:

■ It addresses the gap between a building that does not work as intended and an already over-burdened maintenance and engineering staff.
■ The building is not providing an adequate work environment.
■ There are indoor air quality issues.
■ Mold is present in the building.
■ The building's energy costs are too high compared with similar facilities.
■ Equipment and systems are sustaining damage over the long term from the indoor environment.

■ It's a prerequisite for LEED-EB (Existing Buildings).

Like commissioning, retrocommissioning seeks to identify and correct the root causes of problems, not the symptoms (e.g., "the room is hot"). Also like commissioning, retrocommissioning is concerned with how equipment, systems, and subsystems function together, but it does not generally take a whole-building approach to efficiency.

Retrocommissioning does not include the replacement of significant HVAC and other system components. Rather, it focuses on the verification of the proper controls strategies, sequences of operation,

95

"It is far easier and less expensive to maintain a building that operates correctly than to maintain one that does not."

Rusty Ross

This chapter describes the process for implementing retrocommissioning in new construction and major renovation projects. Chapter 5 covers the commissioning process, Chapter 7 covers the recommissioning process, and Chapter 8 covers continuous commissioning.

In this Chapter

- ◆ Planning Phase
- ◆ Discovery Phase
- ◆ Corrective Phase
- ◆ Project Hand-Off
- ◆ Best Practices

Loose electrical connections or insulation damage may not show up until more equipment increasingly comes on line and electrical loads increase – usually well after building acceptance.

96

control component functionality, operations and maintenance procedures and strategies, and other building optimization opportunities.

Retrocommissioning is not tied to a specific new construction or renovation project and therefore does not follow the same process as commissioning. Sampling is not performed in retrocommissioning. Instead, 100 percent of the applicable components and systems are evaluated.

The four phases that retrocommissioning follows, described in detail in this chapter, are planning, discovery, correction, and hand-off.

PLANNING PHASE

Documentation Requirements:
- ◆ Project Objectives Document
- ◆ Design Intent Document

Identify Project Objectives and Scope

The Owner identifies the objectives for the retrocommissioning project, including in the areas of operational improvement, energy performance, water performance, maintainability, sustainability, indoor environmental quality, and environmental impacts. This will serve as guidance when deciding which equipment and systems are targeted for analysis and potential improvement.

Guiding questions that should be asked to help determine what systems should be retrocommissioned include:

Where have problems consistently occurred?

There is no reason to propagate a recurring and costly headache. System reliability should be a primary focus. The only way to ensure this is to determine and correct forever the root cause of the problem. This knowledge can then be applied to other systems experiencing the same symptoms.

What are the risks of system malfunction?

Consider the risks that failure has on mission, on personnel productivity, on operational production and on other factors such as security, safety, environment, and energy. Consider the impact of redundant and backup systems and parts availability and technical supportability.

What are the political implications of poor system performance?

Consider the purpose and location of the system. Higher priority would probably be given to a fan-coil unit in the Owner's office than it would in a warehouse office space.

How will deficiencies be found without retrocommissioning?

There are several possibilities. However, it usually depends on how much risk you are willing to tolerate and how much cost you can afford. Deficiencies can be found by doing nothing, thereby risking failure at any time and paying higher repair costs when it fails and higher operating and energy costs until that time. Deficiencies can also be found with a proactive maintenance program that combines failure modes and effects analysis and root cause failure analysis with observant preventive maintenance and repair activities. They can also be found through trend analysis of operations, energy, and key performance data.

So consider, for each system being considered for retrocommissioning, their affordability, labor commitment, and degree of risk relative to their commissioning or non-commissioning.

Determine the Commissioning Team

During the planning phase, the Owner will hire an independent, third-party CxA. The CxA will work closely with the Owner, the building Management and O&M staff, and any necessary additional contractors and vendors (e.g., controls vendor).

Retrocommissioning, particularly for less complex, smaller buildings, can also be accomplished in-house with facility O&M staff, although using a

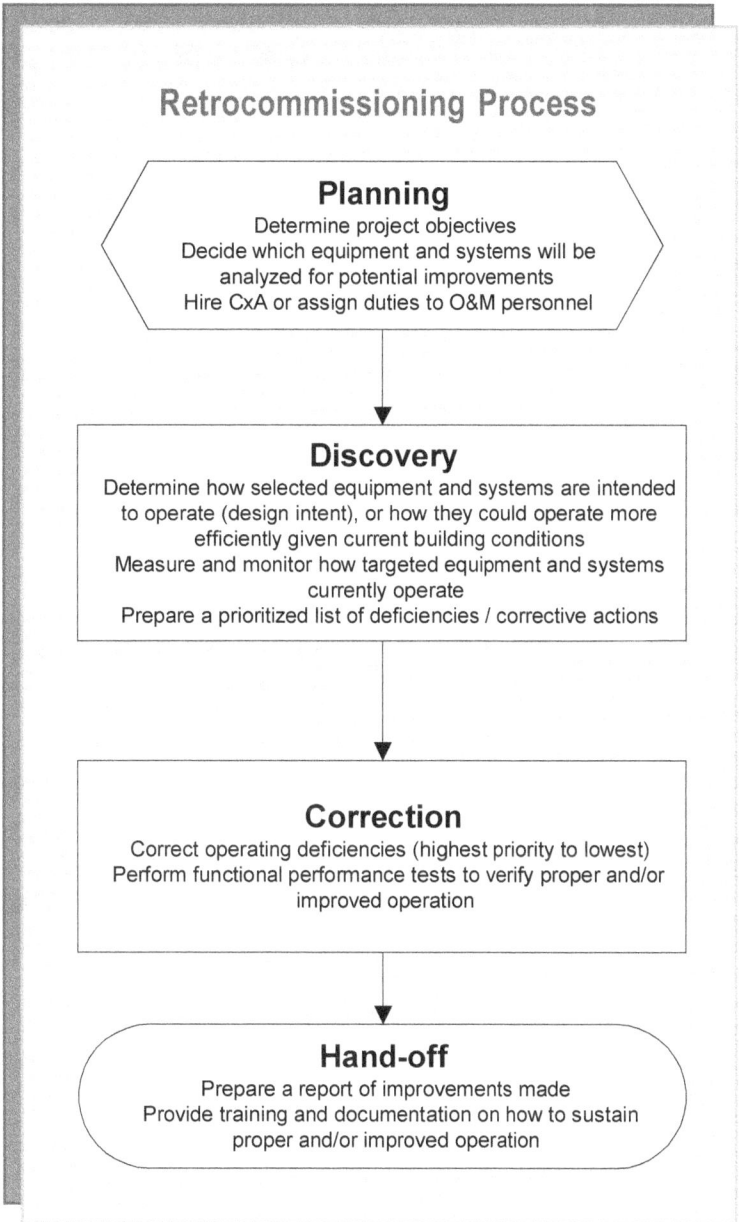

Retrocommissioning Process

Planning
Determine project objectives
Decide which equipment and systems will be analyzed for potential improvements
Hire CxA or assign duties to O&M personnel

Discovery
Determine how selected equipment and systems are intended to operate (design intent), or how they could operate more efficiently given current building conditions
Measure and monitor how targeted equipment and systems currently operate
Prepare a prioritized list of deficiencies / corrective actions

Correction
Correct operating deficiencies (highest priority to lowest)
Perform functional performance tests to verify proper and/or improved operation

Hand-off
Prepare a report of improvements made
Provide training and documentation on how to sustain proper and/or improved operation

97

CxA cuts down significantly on O&M staff workload related to the retrocommissioning process. If choosing this option, the Owner must designate a team of senior members of the facility Management and O&M staff to take on the duties of the CxA.

It is beneficial to involve the O&M staff in retrocommissioning for numerous reasons:

- They know the building systems.
- They have access to all the sections of the buildings.
- They have logs and service records.
- They know the equipment and system Standard Operating Procedures (SOP) that must be followed and personnel protective equipment that must be worn.
- They have access to additional and specialized equipment.
- They know how the control system operates.
- They can make on-the-spot minor repairs.

All involved need to realize, however, that theirs is not a fault-finding mission. Rather, it is to work collectively to optimize and improve the building's efficiency and working conditions.

Retrocommissioning Plan

The CxA develops a retrocommissioning plan that lays out the retro-

commissioning strategy and process that will be followed. This plan identifies the commissioning team members, and is updated and revised as necessary throughout the retrocommissioning project.

The Owner, based on the project objectives and current requirements (which may have changed since the original design intent was developed), draws up a list of equipment and systems that will be targeted for investigation during the retrocommissioning project. The most likely candidates are HVAC, mechanical, and lighting systems, in which deficiencies and problems are common and low-cost improvements can have big impacts.

The following information, as a minimum, should be included in the retrocommissioning plan:

Overview and General Information

- General Building Information
- Definitions and Abbreviations
- Retrocommissioning Objectives
- Purpose of the Retrocommissioning Plan
- Retrocommissioning Scope
- Equipment and Systems to be Investigated for Potential Improvements

Commissioning Team Members

- Points of Contact
- Project Organization Chart

Roles and Responsibilities

- Commissioning Authority (if one is used on the project)
- Owner / Project Manager

The Goal of Retrocommissioning

The goal of new building commissioning should be a 100 percent perfect operating facility. The goal of retrocommissioning is not necessarily a 100 percent perfect facility, since the existing facility may not meet the current operating goals of the occupants. Instead, the goal may be to raise a facility's operating efficiency above the 90-percent mark with little or no extra remedial cost. The remaining 10 percent depends on cost/benefit analysis.

- Building O&M Personnel
- Others

Retrocommissioning Plan
Strategy and Process

- Energy and Operational Analysis and Establishment of Performance Baselines
  - Utility bill analysis and benchmarking
  - Trend analysis and building modeling
  - Documenting master list of findings
  - Energy and operational cost savings estimating
- Document Review
  - Available construction documents
  - O&M Manuals
- Staff Interviews
  - Maintenance Staff
  - Occupants
- Implementation
  - Implementing O&M and Capital Improvements
  - System Documentation (Control Sequences)
  - O&M and Systems Manuals
- Functional Performance Testing
  - Overview and Process
  - Sampling Strategy
  - Deficiencies and Retesting
- Monitoring and Verification
  - Monitoring System Performance
  - Measuring Energy Savings
- Project Final Report

Written Work Products

- List of deliverables

Document Collection

The retrocommissioning team collects general facility and equipment and system-specific data during this phase. With the caveat that many of the original documents may not exist or may be out-of-date, the following documents should be collected and made available, if possible, to the retrocommissioning team:

- Available construction documents, including plans, inspection reports, and O&M manuals.
- Equipment baseline data.
- Utility bills (previous 12 to 24 months) and energy data.
- Equipment inventory, including size, capacity, and age data.
- Equipment maintenance information.
- Control sequences of operation.
- TAB reports.
- Equipment or operations benchmark, key performance indicator, and trending information.

DISCOVERY PHASE

Documentation Requirements:
- Retrocommissioning Plan
- Master List of Deficiencies and Improvements
- List of Improvements Selected for Implementation

During the investigation phase, the CxA works with the facility management and O&M staff to investigate the building and its equipment and systems. The result of this investigation is a prioritized list of im-

All involved need to realize that theirs is not a fault-finding mission. Rather, it is to work collectively to optimize and improve the building's efficiency and working conditions.

99

Case Study: Retrocommissioning in Action

The air conditioning system at a converted (from a shopping center facility) telephone call center with 250 operators was not providing sufficient comfort. The cost for a new air conditioning system was $250,000. The Owner had called in an air-conditioning contractor, mechanical engineer, and controls contractor to correct the problem, and each was unsuccessful. Finally, during retro-commissioning it was discovered that return air grilles were missing from the newly renovated facility and the building was starving for return air. The air-conditioning contractor, mechanical engineer, and controls contractor failed because they focused on their specific areas of expertise and overlooked the simple.

provements to make during the implementation phase.

Document Review

When implementing retrocommissioning, the original design intent or up-to-date construction documents (particularly drawings) may not be available. Even if they are, the Owner's requirements may have changed since these documents were originally prepared due to reasons such as mission changes, occupancy reassignments, installation or removal of interior walls and partitions, and introduction of new technologies. If these documents exist, it is essential that the retrocommissioning team review them for comparison with the Owner's current requirements and existing conditions to determine if the original size and capacity units can provide the current required output.

The CxA reviews the documentation developed and/or compiled during the planning phase. An analysis is conducted to look for signs of equipment and system deterioration or sub-optimal performance. Maintenance histories and operations trends relative to baseline information will be indicative of problems that may have root causes initiated with system design, maintenance procedures, personnel qualifications and training, scheduling, changed conditions resulting in over- or under-sizing, controls, component failure, or other deficiencies.

Review and analysis of energy data and utility bills likewise indicate problematic trends, as well as opportunities where both energy and energy cost savings may be achieved.

Personnel Interviews

Personnel interviews with at least two specific groups of personnel are invaluable to the retrocommissioning team during the investigation phase – the maintenance staff and the building occupants.

Interviews with the maintenance staff will help determine their perception of current problems that may be associated with the equipment and systems, the facility O&M culture, O&M practices, or other influential factor on system efficiency and reliability. The interviews will also help the retrocommissioning team better understand operating strategies and equipment condition.

Records of the interviews should be maintained for future reference. The interviewers need to be aware that the information may be factual, hearsay, perceived, or even political in nature. Look for patterns.

Similarly, interviews with building occupants will help determine their perception of current problems that may be associated with equipment and systems, facility design, Owner response, and the operations and maintenance culture. Ask if they hear "strange sounds" or "noise" from the mechanical and ventilation systems. Look for patterns.

As before, interview records should be maintained and the interviewers need to be attuned to the fact that the information may be factual, hearsay, perception, or politically-influenced.

Site Assessment

According to NFPA Standard 70B, "as soon as new (electrical) equipment is installed, a process of normal deterioration begins. Unchecked, the deterioration process can cause malfunction or an electrical failure." The same holds true for mechanical systems.

The CxA conducts a site assessment to evaluate how and why building equipment and systems are currently operated and maintained. The CxA identifies for further investigation any significant problems reported by the O&M staff and building occupants. The site assessment for retrocommissioning also includes a facility survey to determine occupancy and space utiliza-

tion. The site assessment addresses the following major issues:

- Building occupancy and space utilization.
- System and equipment condition.
- Overall building energy use and demand.
- Areas of highest energy use and demand.
- Utility bill analysis and benchmarking (bills from previous 12 to 24 months).
- Air and water flow rates, calibrations, and flow coefficients.
- Actual control sequences for each piece of equipment and each system included in the project and their functionality.
- Equipment nameplate information.
- Equipment maintenance approaches and issues.
- Facility zone temperature and humidity levels.
- Facility lighting and CO_2 levels
- All significant control and operational problems.
- All significant occupant comfort problems.

101

Effective Interviewing

- ◆ Do not readily accept the first thing anybody tells you – they will be telling you symptoms or results.
- ◆ Ask the same question three different ways to drill down beyond their concern to the root cause.
- ◆ Be patient.
- ◆ Allow for silence – people are uncomfortable with silence and will begin talking.

- Locations of building trouble spots.
- Current O&M practices.

Diagnostic Monitoring and Testing

Diagnostic monitoring and testing provides information on temperatures, critical flows, pressures, speeds, and currents under typical operating conditions. By analyzing this information, the CxA determines whether the systems are operating correctly and in the most efficient manner. The following diagnostic methods are commonly used: energy management control system (EMCS) trend logging, stand-alone portable data logging, and manual functional performance testing.

The CxA schedules the implementation of the diagnostic monitoring and testing. The CxA works with the O&M staff to make sure equipment and systems are ready for testing. It will be necessary for the Control technician to assist with the EMCS trend logging.

The CxA oversees functional performance testing, which is performed by the O&M staff. Functional testing is the dynamic testing of systems (rather than just components) under full operation. For example, the chiller pump would be tested interactively with the chiller to see if the pump ramps up and down to maintain the differential pressure set point. Systems are tested under various modes - low cooling or heating loads, high loads,

component failures, unoccupied, varying outside air temperatures, fire alarm, power failure, etc. The systems are run through all of the control system's sequences of operation. Components are verified to be responsive per the prescribed sequences.

The findings of diagnostic monitoring and testing are analyzed and compared with the site assessment data. Any resultant changes are added to the Master List of Deficiencies and Improvements.

Master List of Deficiencies and Improvements

The survey information is reviewed by the CxA. Based on the input from the site assessment, the CxA develops corrective strategies and prepares a Master List of Deficiencies and Improvements. This forms the basis of the project decision making and problem prioritizing process. Specific solutions to the problems found are identified.

Every problem, deficiency, or opportunity for increased efficiency that is found during the investigation phase is summarized on the Master List, including any minor adjustments and repairs that are made during the course of the investigation process. The list should include, for each identified issue, the name of the affected equipment or system, a description of the deficiency or problem, and recommended action(s).

Control systems have become so sophisticated that few end users really understand fully how to use them to optimize system performance. Typically, they are not programmed or calibrated correctly on the front end, and maintenance professionals often bypass them completely to address exigencies.

In addition, any opportunities for the installation of energy conservation measures or adoption of more efficient procedures and practices observed during the investigation process, related to the retrocommissioned systems or not, are also be identified.

In the final report, the CxA organizes the deficiencies on the Master List into categories, such as design, installation, maintenance, and operation. This allows the Owner to identify trends that are contributing to problems in certain areas.

The retrocommissioning team also performs "quick fixes" during the discovery phase. These are simple repairs and adjustments, such as connecting an unconnected section of flex duct, unblocking a duct, tightening a fan belt, and cleaning a coil. Though relatively insignificant in nature, these may be masking a real problem.

Corrective Phase

Documentation Requirement:
◆ Master List of Deficiencies and Improvements

During this phase, cost-effective opportunities are selected for implementation. Based on the findings of the site assessment and diagnostic monitoring and testing, and using the Master List as a guide, the Owner determines which recommended improvements to implement. Items should be prioritized according to cost effectiveness, criticality, and how effectively they meet the project objectives. It is likely that many of the recommendations will have no cost/low-cost solutions, such as personnel habits, that can be implemented right away.

To aid in the decision making, the CxA provides economic analyses (estimated cost, savings, payback, and return on investment) on those items that can be quantified.

If substantial modifications or capital improvements are required, formal commissioning of the affected systems should be included during the various construction phases as described in Chapter 5.

Project Hand-Off

Documentation Requirements:
◆ Updated Building Documentation
◆ Final Retrocommissioning Report

During project hand-off, equipment and systems are tested again to confirm the operational and energy performance of the installed improvements.

103

Monitoring and Verification

Diagnostic monitoring and testing is performed again after the recommendations have been implemented. Post-implementation data is compared to pre-implementation data to confirm that the improvements are integrated and working properly together and have the desired effect on building performance.

Diagnostic monitoring and testing is also used to benchmark the performance of the improvements. This establishes parameters for measuring the performance of the improvements throughout the life of the equipment and systems.

Update Building Documentation

The following building documentation is updated by the Owner or by the CxA, depending on the scope of the service agreement, to reflect changes made to equipment and systems during the retrocommissioning process:

- One line drawing schematics for each system affected by the improvements.
- Operation and maintenance manuals and system operations manuals (including updated sequences of operation for equipment).
- Energy management plan, including guideline for implementing an energy accounting and tracking system with performance benchmarks.
- Preventive maintenance plan, or a guideline to implementing a preventive maintenance plan if one does not exist.
- New condition baselines established for the maintenance program.
- Operation and maintenance staff training materials, including list of operational strategies and operational assessment.

O&M Staff Training

If a building is overly sophisticated, the staff's maintenance capabilities need to be aligned with that sophistication.

The CxA will provide additional training, or arrange for training with appropriate vendors, if warranted by the improvements made to the equipment and systems.

Final Retrocommissioning Report

The final retrocommissioning report includes a summary report of participants and their roles, building description, project objectives, an overview of the retrocommissioning scope, and a general description of testing methods. The following items typically are also included:

Heating is becoming a big issue in HVAC. Lighting heat load historically has been designed into HVAC load calculations. Now with efficient lighting, there is a problem getting some spaces warm enough.

104

"It has been estimated that 70- to 80-percent of all unplanned shutdowns (of electrical systems) are due to human error, meaning that only 20- to 30-percent of unplanned shutdowns are due to equipment malfunctions or poor design." Source: *A Practical Guide for Electrical Reliability*, EC&M (Oct 2004).

- Retrocommissioning plan
- Master List of Deficiencies and Improvements
- Cost/savings analyses for each implemented recommendation
- Site assessment results
- Diagnostic monitoring and testing results (pre- and post-implementation)
- Controls sequences and block diagrams indicating component functions and relationships
- Updated TAB data
- Equipment condition baseline data
- All completed functional performance test forms
- Recommended frequency for recommissioning
- Recommended frequency for recalibration of sensors and actuators
- Documentation of implemented recommendations
- Energy saving features and strategies used in the building
- Listing of all user adjustable set points and reset schedules
- Recommended frequency for review of set points and reset schedules
- Photographs of every deficiency found

Turning the Report into Action

It is crucial for the facilities team to buy into the report and results for it to succeed, then for the documentation to be used and maintained. It is good practice to develop a checklist of action items that can be addressed, confirmed, filed, and referenced.

BEST PRACTICES

- Keep the retrocommissioning scope focused on the facility's mission.

- Concentrate on only the essentials.

- Collect only important and especially useful information.

- Link the retrocommissioning to the facility's operating and business objectives:

  - It documents performance criteria and data for tracking, evaluating, and improving systems.

  - It correlates system performance to monetary indices through energy savings, improved maintenance, improved worker productivity, and occupant satisfaction.

- The documentation required by each commissioning process can be daunting, but will prove an invaluable resource for building management, O&M, and future commissioning efforts.

Over the span of a few years, it is highly likely that the building systems have been modified. While modifying systems is not a problem in itself, the lack of documentation can be.

105

STUDY QUESTIONS

1. What are the documentation requirements for each phase of a retrocommissioning project? Why are these documentation requirements so important to the overall success of the retrocommissioning project?

2. What system and equipment are the best candidates for retrocommissioning investigation, and why?

3. How do the steps involved in retrocommissioning differ from the steps involved in new construction commissioning?

4. If building documentation already exists, is there a need to further develop or refine documentation during retrocommissioning of that building?

5. What are the methods of investigation used during the retrocommissioning discovery phase?

6. What documents make up the final commissioning report developed at the conclusion of the retrocommissioning project?

7. What role does the O&M staff play in retrocommissioning? The Owner? Is it necessary to hire an independent CxA?

Chapter 7
Recommissioning Process

Recommissioning refers to commissioning of an existing building that has already gone through the commissioning process. It is performed long after the facility is constructed and placed into service. Some sources and providers use the term "recommissioning" synonymously with "retrocommissioning" to collectively address the commissioning of existing buildings.

Like all other forms of commissioning, the goal of recommissioning is to ensure that all power-using and power-conserving systems in a building work together to meet the needs of the current occupants and the actual performance requirements of the Owner.

Recommissioning provides additional opportunities to improve facility efficiency and addresses issues that may have arisen since the original commissioning. It can help reduce energy consumption, maximize the efficiency and output of the air and water distribution systems, enhance performance, and enhance the occupants' working environment and comfort. Like commissioning, recommissioning may involve functional performance testing of most or all major building systems,

particularly if they have been problematic or highly energy inefficient. However, recommissioning is most often applied to the existing building's HVAC, refrigeration, and electrical systems and their controls, which often are the sources of the biggest operational problems:

■ Increased occupant complaints
■ Increased or fluctuating energy use
■ Increased maintenance calls

Corrected, these systems also are likely to produce the biggest cost savings. Recommissioning provides a systematic approach for finding,

"Perserverance is the hard work you do after you get tired of doing the hard work you already did."

Newt Gingrich

This chapter describes the process for implementing recommissioning in existing buildings. Chapter 5 covers the commissioning process, Chapter 6 covers the retrocommissioning process, and Chapter 8 covers continuous commissioning.

In this Chapter

- ◆ Project Planning
- ◆ Design Review
- ◆ Implementation and Verification
- ◆ Periodic Review
- ◆ Best Practices

108

Recommissioning every five years is required of all State of California buildings larger than 50,000 square feet.

verifying, documenting, and correcting their deficiencies.

Recommissioning can be undertaken as an independent process in response to a specific requirement or concern, or periodically scheduled as part of the building's operations and maintenance program. Systems tend to shift from their as-installed conditions over time due to normal wear, user requests, and facility modifications.

Since a considerable investment has already been made in the initial commissioning of the facility, it is recommended that it be recommissioned about every three to five years, as scheduled and as part of the preventive maintenance (PM) protocol. One-third to one-fifth of the building's applicable systems could be recommissioned each year on a rotating basis. In that way, recommissioning is cost- and personnel-budgeted as "business as usual" and is not a special event.

In general, the more substantial changes that a facility goes through, the more often it should be recommissioned if a continuous commissioning program is not in place.

Preventive maintenance in itself is not enough because its procedures tend to focus on specific component care and not on an integrated system. The *operations* side of O&M involves observing and monitoring the building's systems to determine how and when they operate and if they are producing and delivering the desired result. However, in the typical facility, other priorities take precedence on the time-strapped staff and eventually even this good intention may become neglected beyond periodic gauge readings and monitoring of energy management system output parameters. That is, there is little-to-no trending or analysis conducted that could be indicative of sub-optimal conditions or emerging problems.

During recommissioning, the tests that were performed during the original commissioning are repeated to check for persistence in results against the baselines established at that time. The objective is to ensure that the building is operating as designed or according to newer operating requirements.

The development of new project documentation and testing procedures and forms is not required. However, these documents can be updated if the building, its systems, equipment, and mission and occupancy habits have changed dramatically since their initial commissioning. Nameplate data, inventory lists, training curriculum, and as-built drawings should be updated in any event, as part of the maintenance program.

Recommissioning generally includes:

- Communicating among the commissioning team and facility staff, occupants, and users to uncover building system problems and opportunities.
- Establishing that the original basis of design (if available) and operation plan are still appropriate for the use, occupancy, and occupant mission of the building.
- Reviewing and benchmarking key systems operations/performance against the previous commissioning recommendations and baselines.
- Identifying past and current persistent problems.
- Reviewing maintenance program procedures, schedules, and protocols and verifying their effectiveness against actual equipment observations.
- Reviewing the operators' and maintainers' technical capabilities relative to the building's needs and to the training program (or contractual requirements).
- Performing energy analysis of available data, including utility bills.
- Identifying specific energy conservation measures, particularly no-cost and low-cost solutions.
- Performing a condition assessment using condition monitoring technologies.
- Recommending repairs and modifications to optimize building performance.

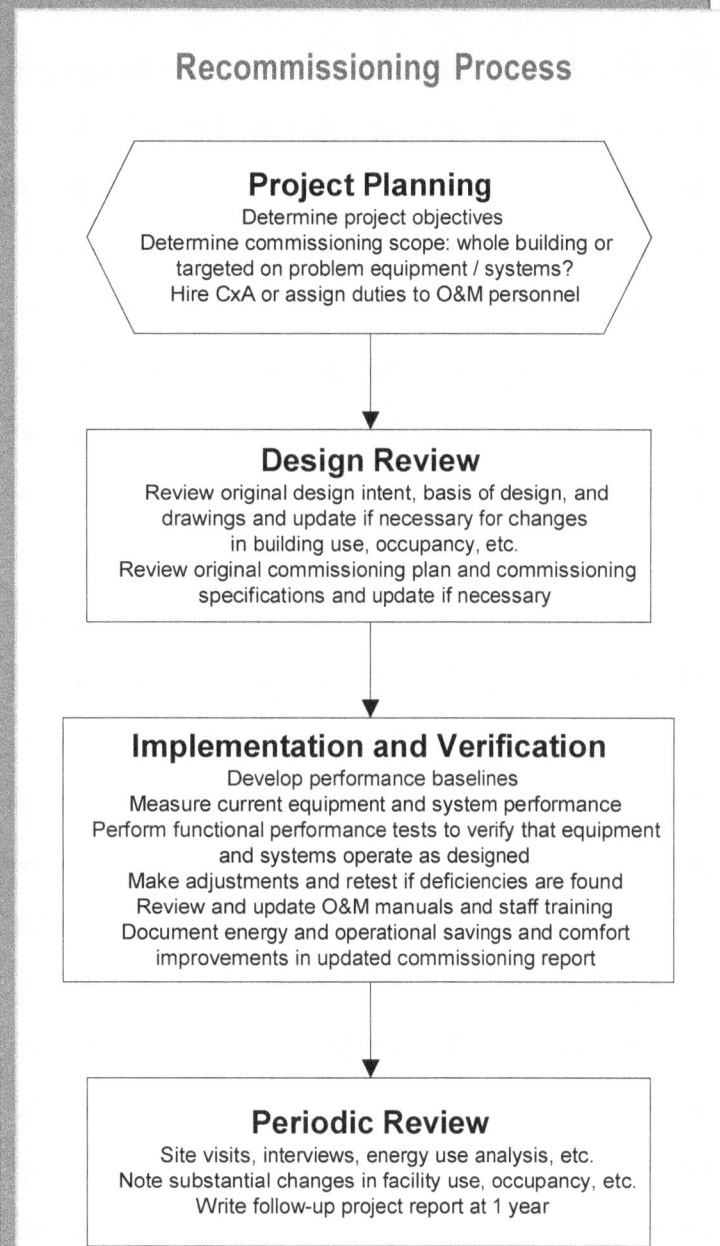

- Validating and/or modifying the operations/controls sequencing as appropriate for optimum operations.
- Conducting testing, adjusting, and balancing (TAB) on the HVAC air and water distribution systems, as required.

Recommissioning is *not* a redesign process.

PROJECT PLANNING

Documentation Requirements:
- Project Objectives Document
- Original and Revised Commissioning Reports

During project planning, the Owner determines the project objectives and recommissioning scope. The Owner determines the recommissioning team, including hiring an independent, third party CxA or assigning CxA duties in-house.

Project Objectives Document

The Project Objectives document is prepared by the Owner. It provides a general introduction of why the building is being recommissioned, and typically lists concerns, problems, and occupant complaints. It also identifies requirements relative to energy performance, water performance, maintainability, sustainability, indoor environmental quality, and environmental impacts.

The Owner's operating requirements are also an important part of this document. For example, the need to maintain a strict range of humidity levels may require an air handling system to stay "on" at all times, when otherwise it could be shut down during nights and weekends.

If a specific system is identified, this document will include a list of components that require functional testing. For example, if the cooling system is identified as problematic, all system components would be listed for functional testing: chiller condenser, chilled water pumps, strainers, water treatment, heat exchanger, and controls.

The Project Objectives document is used by the commissioning team as a guide and to stay focused when conducting the recommissioning tasks and in updating the building documentation.

Determine the Recommissioning Team

The recommissioning team consists of the Owner, CxA, Facility Management, O&M staff, any necessary additional contractors (such as for controls and TAB support), and vendors. An independent CxA can be hired to perform recommissioning, or the function can be assigned in-house, using previous commissioning

---

Common Recommissioning Findings

- Variable speed drives no longer modulate properly.
- Controls are manually overridden.
- Improper sequences of operation result in inefficient or ineffective machine operation.
- Building configuration, use, and/or occupancy have changed without considering the impact on HVAC and other systems.
- Equipment is inadequately maintained, and observation of its condition does not necessarily match what is indicated on the PM checklist.
- Technician training has not kept pace with the fast-paced, constantly evolving technology.

---

guidance and data, existing test forms, and knowledgeable staff.

In either case, a work plan must be established that identifies required support and training. The work plan includes a work schedule with milestones and target dates. If the work is done in-house, the staff should be dedicated to the effort through its completion with no or minimal regularly-assigned O&M work duties and interruptions.

Kick-off Meeting

The project kick-off meeting brings the recommissioning team together to review the Commissioning Plan. At a minimum, the CxA, TAB provider and controls technician (as applicable), Owner and/or designated representatives, and O&M Manager should attend. Each team member is introduced and their responsibilities are stated. The Owner's expectations, goals, and objectives are reiterated. Communication protocols, schedules, and safety issues are discussed.

The kick-off meeting is also a good opportunity to conduct an interview of occupants, users, operators, and maintainers to determine the scope and degree of problematic conditions from their perspectives (are the persistent hot and cold calls a systematic problem). The interview survey may take any form, but should follow a prescribed outline. Sample interview questionnaires for building and maintenance Management are included in

Appendix A.

Procedures

Unlike commissioning, recommissioning involves testing and inspection that occurs in occupied spaces. This can have significant impact during testing, since operating sequences need to be tested and the building's working environment conditions will be changed temporarily. Access to equipment in occupied spaces will also be a challenge, particularly if ladders and access to overhead units, air distribution components, and lighting fixtures are required. The occupants need to be informed and brought into the process to minimize disruptions to themselves and to the testing.

Document Review

The document review includes reviewing work orders and trouble calls from the building Owner, occupants, and users to identify and survey recurring, persistent, and/or serious problems. Comments such as hearing "loud noises" from ductwork or feeling walls vibrate whenever an AHU kicks on are invaluable in determining root causes of problems.

The commissioning team studies the building's documented problems and looks for trends and links. Maintenance records, operating cost records, and energy use information, as available, will prove helpful.

As-built drawings are reviewed to understand system configurations and modifications from the original design documents. The building history is also reviewed to check on

Building occupants will modify their personal space to achieve comfort. Tampering with thermostats, sensors, and diffusers, and introducing devices such as portable heaters, contribute to the inefficiency of a building's HVAC systems.

111

Case Study: No-Cost Measures Can Bring Big Savings

The building management system (EMS) of a corporate office complex in Massachusetts was programmed to allow after-hours employees to dial a code to turn on a small (2,000 SF) area of lights as needed. However, over time the dial-in codes were misplaced, so security staff would manually turn on the lights for an entire floor. Typically, this meant that two floors of lights (216,000 SF) would be illuminated for perhaps five night owls.

To address this problem, the dial-in codes were redistributed to all staff members and posted in their appropriate zones. This no-cost measure resulted in annual cost savings of about $45,000. (Source: Haasl, T. et al, Pacific Northwest National Laboratory, *Retro-Commissioning's Greatest Hits*, March 2000.)

112

During the design review phase, the commissioning team reviews the original design and subsequent commissioning documentation to update and reflect, if necessary, changes or concerns such as:

- Different use or occupancy patterns of the facility since its previous commissioning.
- How people interact to operate the building systems from day to day.
- Outmoded equipment or systems.
- Equipment or systems that do not operate optimally and have high failure rates.
- Persistent high energy costs.
- Occupant complaints.
- Need for the building to achieve greater energy and water savings and a healthier indoor environment (e.g., for LEED-EB certification).

Unfortunately, original information sources, such as calculations and equipment specifications and submittals, may no longer be available and may need to be developed.

The CxA reviews changes made to the design documentation for commissioning facilitation, energy efficiency, control system optimization and strategies, operations and maintenance effectiveness, indoor environmental quality, O&M documentation, training, and potential enhancements to the mechanical and electrical systems. This creates the benchmark against which all inspection and test data is compared.

impacts caused by changes to the internal configuration (new partitions?), occupancy (now 150 occupants in the space, up from 75?), and mission (former warehouse converted into a laboratory or IT Center?). O&M manuals are reviewed to ensure that product descriptions and recommended maintenance schedules and procedures are accurate for the equipment now in place. Good O&M manuals include only applicable O&M guidance and data specific to the equipment installed – *not* non-applicable models or manufacturer advertising.

DESIGN REVIEW

Documentation Requirements:
- Updated Design Documentation
- Updated Commissioning Plan

Commissioning Plan

The changes made to the design documentation necessitates updating the most recent Commissioning Plan as well. The Commissioning Plan is updated to include information specific to the current recommissioning: commissioning objectives, identification of equipment and systems being commissioned, commissioning team members, project organization, current conditions and changes since the previous commissioning, and apparent and declared problems and opportunities that need further verification and investigation. Areas of emphasis may include:

- Heat exchangers
- Heating and cooling systems
- Air delivery and ventilation systems
- Control systems
- Air and water system testing, adjusting, and balancing (TAB)
- Lighting
- Building envelope

Recommendations focus on optimizing the performance of equipment and systems in place. Upgrades to include new technologies and energy efficient measures may be recommended resulting from the recommissioning effort, but are treated as a separate action.

IMPLEMENTATION AND VERIFICATION

Documentation Requirements:
- Commissioning Report
- Commissioning Plan (Final)

- Functional Performance Checklists

During the implementation and verification phase, the CxA measures current equipment and systems performance and develops performance baselines. Functional performance tests are performed to verify that equipment and systems operate as required. Selected improvements are made, and improvements and operational and energy savings are verified and documented. The CxA also reviews and recommends (or performs, as applicable) updates to O&M manuals and staff training.

Performance Baselines and Equipment and Systems Measurements

Baseline energy models of building performance are established to document a performance profile before and after recommissioning. Baseline energy models are established using anyone or combination of the following data:

- Short-term measured data from dataloggers, building automation system (BAS), or energy management and control system (EMCS).
- Metered long-term hourly or is-minute whole building energy data.
- Historic utility bills for electricity, gas, steam, and chilled or hot water.

Trending allows the recommissioning team to observe performance and output parameters under vari-

113

ous modes and conditions over time. Variables often trended include energy, temperatures, pressures, flow rates, weather data, and set points. By analyzing trends, the recommissioning team characterizes performance and validates whether or not the systems operate correctly.

Trending may be conducted by reviewing and plotting BAS and EMCS data over time. The commissioning provider may require assistance from the controls vendor, particularly in developing, downloading, and formatting system data for analysis.

Dataloggers also provide a trending capability. These are small, portable, battery-operated devices utilize software that can be downloaded onto a laptop and easily graphed and analyzed. Data collected typically includes temperatures, humidity, pressure, current, light levels and durations, and similar information.

A utility bill analysis, conducted early in the project, gives the commissioning team a good understanding of how the building is consuming en ergy and the direction toward which they should target their efforts.

All major equipment and systems are checked to record current operating parameters and to verify the correct operation of their parts. This is analogous to the verification checks in new construction and renovation commissioning. Verification checklists that were developed when the building was first commissioned can be used again.

This is also an appropriate time to check the equipment condition, if it is not already done in a Reliability Centered Maintenance (RCM) program, using available condition and predictive technologies such as infrared thermography, vibration analysis, equipment alignment, airborne ultrasonic leak detection, oil analysis, insulation testing, and motor analysis.

Any significant problems found during equipment and systems measurements are noted and corrected by the O&M staff prior to functional performance testing.

Sensors and actuators should be calibrated prior to functional testing, and checked by the CxA.

114

How *not* to repair a leak (actual example taken in the field).

Functional Performance Testing

The CxA schedules, oversees, witnesses, and documents the functional performance testing of all equipment and systems according to the Commissioning Plan. The facility O&M staff executes the tests, which use the same protocols and forms that were developed when the building was last commissioned. Functional testing includes operating the system and components through the significant normal and emergency modes of operation, including:

1. Each of the written sequences of operation
2. Start-up and shut-down
3. Unoccupied mode
4. Manual mode
5. Staging
6. Miscellaneous alarms
7. Power failure
8. Interlocks with other systems or equipment

A sample functional performance test form is included at the end of *Appendix A, Sample Commissioning Forms*.

Simple repairs and adjustments, such as belt replacement, damper adjustment, and sensor calibration, not detected earlier and found during the investigation may be done at this time to maximize system efficiency and to enable system testing, adjusting, and balancing.

Functional testing is accomplished using a combination of conventional manual methods, control system trend logs, and stand-alone data loggers to provide a high level of confidence in proper system function.

Recommissioning In Action

A middle school was experiencing severe indoor air quality problems and the school district requested the help of a commissioning consultant. Recommissioning found that among other things, all of the building's fresh air vents were closed and systems were not operating as originally intended. Since the original controls contractor had gone out of business, each time an occupant complained about the building's comfort level over the years, the maintenance staff responded by closing the offensive vent, thereby addressing the symptom, not the root cause.

The commissioning team confirmed that the HVAC systems, when new, had been installed correctly. The facility would have most likely functioned adequately had the support staff understood from the start the control system, its maintenance requirements, and how it functions. (Source: E. Thomas Lillie, "Better Late than Never," *Engineered Systems Magazine*, May 25, 2000.)

Building Management Systems (BMS)

Recommissioning the BMS is one of the most cost effective building and energy performance measures that can be undertaken. The operational status and correct programming of each strategy of the BMS is verified. This requires trending of the BMS data over time to see if the strategies are actually controlling the equipment as they should. The cause of malfunctioning control strategies is then determined. For example, is the system not operating' correctly because of a software programming error? Overridden control strategy? Ineffective sensor locations? Malfunctioning components?

Countdown of the Seven Most Common BMS Problems

7 – The BMS operator does not have a good understanding of energy conservation.

6 – Failed components are not replaced.

5 – The full sequence of controls is not verified at the time of system acceptance.

4 – Control devices and sensors are disabled "temporarily" to satisfy immediate occupant complaints.

3 – Disabled energy conservation features, such as by power outage, are never reset.

2 – The BMS is not properly programmed.

1 – The maintenance staff does not have the time, or is inadequately trained, to thoroughly diagnose BMS problems.

All analog and digital in- and out-points are verified. Heating and cooling setpoints and setbacks are checked, and heating and cooling reset schedules are checked for accuracy.

Testing, Adjusting, and Balancing (TAB)

Testing, adjusting, and balancing may or may not be part of the recommissioning process. TAB is the process of adjusting HVAC system components to supply air and water flows to match the requirements. Inordinate indoor temperature fluctuations, excessive drafts, and improper air distribution causing hot and cold spots lead to occupants' discomfort and excessive energy use. Indications that TAB is required include frequent occupant complaints of discomfort due to hot and cold spots; cardboard taped over air registers at workstations; a significant change in

the building configuration, use, and/or occupancy; and the need for the maintenance staff to frequently adjust the HVAC components.

A TAB analysis involves the measurement and verification of air system flow rates, water system (HVAC) flow rates, temperatures of heating and cooling delivery systems, positions and functionality of delivery system flow control devices, settings and operation of the controls system and components, and fan and pump speeds and pressures. Observations are made against the building's design documentation and are documented in a formal Test and Balance Report.

Maintenance Program

In existing facilities, a review of the maintenance practices and observation of the actual equipment appearance and operation may uncover deficiencies in the program. Recommissioning provides an opportunity to review the program, identify deficiencies, recommend improvements, establish best practices, identify training needs, and document a new program, as required.

Deficiencies and Retesting

Each equipment and system tested must pass its functional performance test to ensure that it is installed and operating according to the current requirements. Corrections of minor deficiencies (e.g., fixing a controller, adjusting alignment) identified can be made by the O&M staff during the tests at the discretion of the CxA.

The CxA records the results of the test on the procedure or test form. Deficiencies or non-conformance issues are noted and reported to the Owner on the Master List of Findings or Findings Log. This is perhaps the most important tool of the recommissioning process and assists the Owner in prioritizing recommended and needed corrective measures. The list identifies every finding detected, including fixes that are made in the field, estimated costs, savings, and paybacks. The list may appear in any order, but categories such as no cost/low cost, medium cost/ short term, and high cost/long term have proven to be valuable for prioritizing, budgeting, and project programmmg.

Systems are retested after the deficiencies are corrected to ensure that they do, in fact, operate and deliver the desired outcome as needed.

Implementation of Improvements

Many changes that are made will be no- or low-cost. For recommended major improvements, such as replacing an old boiler system with an energy-efficient modulating type with reheat or installing a new building automation system, information on costs, estimated energy savings, payback, and return on investment are prepared. In addition, an A/E and specialized contractors may need to be consulted for complicated improvements.

The highest priority is to solve existing problems with no- and low-cost measures. Typical examples include:

- Calibration of building controls, such as thermostats and occupancy sensors.
- Adjusting BMS schedules to ensure that equipment runs only when necessary.
- Checking for and replacing malfunctioning steam traps.
- Cleaning condenser, evaporator, and boiler heat exchanger tubes.

Implementing higher cost corrective measures and non-remedial improvements that optimize equipment and systems follows. The Owner's approach to implementing improvements depends on in-house capabilities, life-cycle cost analysis, available funding, and the Owner's degree of comfort with the recommendations relative to the impact on operations and mission.

Once the improvement is completed, the system documentation is updated to reflect changes to affected control sequences. Also, it is important to retest the equipment and systems to ensure that the improvements are working as expected. The retesting can take the form of utility trending, observation, datalogging, functional testing, or as a

117

Case Study - Finding the Optimal Solution

The recommissioning of a 300,000 SF hospital in California indicated that triple duty valves on the building's condenser pumps were only 20% open. They were throttled back because the pumps were significantly oversized and pumping too much water. Throttled valves reduce water flow, but also add pressure drop to the system, thereby wasting energy.

In most cases, water flow is best reduced by trimming the pump impellers and opening the valves at the pump discharge. If pumps (and fans) are equipped with variable speed drives (VSD), it is tempting to balance the system by slowing the pump down with a drive rather than trimming the impellers. While better than throttling, the result is not optimal since drive efficiency drops as a function of load and drive speed.

A recommissioning recommendation may be to best optimize the system's overall efficiency by adjusting the impeller size so the pump delivers the design flow when the drive is at full speed, and then using the VSD to match actual load conditions. (Source: Haasl, T. et al, Pacific Northwest National Laboratory, *Retro-Commissioning's Greatest Hits*, March 2000.)

combination of these, and the data compared to the initial baseline data. This new baseline data becomes the baseline against which future recommissioning activities will be compared.

Monitoring and Verification

The impact of the minor adjustments made during the functional performance testing and of any implemented improvements is monitored over a reasonable time to verify their impact on occupant comfort, on operational and energy performance, and on control sequences and schedules.

Room-by-room measurements are made using hand-held meters or portable dataloggers. Energy model data is collected using the short-term methods discussed previously. Utility bills and meter data for electricity,

gas, and chilled and hot water are reviewed for long-term monitoring.

Final Commissioning Plan

The CxA finalizes the updated Commissioning Plan after functional performance testing is complete, recommended improvements have been implemented, and there are no more changes or equipment substitutions planned.

Commissioning Report

The fmal Commissioning Report is a record of the recommissioning activities and measures implemented and recommended for implementation. It incorporates a summary report of participants and their roles, building description, project objectives, an overview of the commissioning scope, and a general description of testing methods and results. The Commissioning Report includes and is a formal, permanent record of the following:

- Executive summary.
- Commissioning Plan (updated).
- Current building performance.
- Description of existing system conditions.
- Functional performance test results, including a list of deficiencies found and corrected.
- Description of implemented improvements and their impact on system performance.
- Description of implemented operation and control procedures and their impact on system performance.
- Cost/savings analyses for each implemented recommendation.
- Modifications to O&M practices and guidelines.
- Monitoring and verification results.
- Recommended frequency for recommissioning.
- Additional recommendations.
- Estimated energy savings in MBtu/SF and $/SF.
- Estimated cost and payback calculations for recommendations not implemented at the time of the recommissioning.

It is very important for the Owner and facility staff to receive adequate documentation and training to enable them to make the recommended improvements. The Commissioning Report becomes new documentation for the building's historical files and for implementing, operating, and maintaining the measures and prescribed operating parameters.

Operation and Maintenance Manuals

As a function of the recommissioning effort and continuous improvement, the CxA reviews the operation and maintenance manuals and updates the manuals as needed. Particular attention is given to revised as-built drawings, design intent modifications, changed sequences of operation, new and updated operating instructions, functional performance test procedures and results, new condition baselines, guidelines for continuous maintenance, improved maintenance schedules and checklists, updated manufacturer and vendor contact lists, training requirements, and any new warranty information. The manuals are also reviewed mindful of any changes to the system function and to user and occupancy patterns.

Equipment and Controls Documentation

Similarly, inventory lists are updated for the building's main energy consuming equipment. Important information typically includes nameplate data, equipment identification and location, and the date installed. Updated control system documentation should include a full and complete point

119

Sources for Help

Some Federal and state agencies maintain electronic libraries available through the Internet that can assist the user in recommissioning their facilities. These libraries include tools, specifications, regulations, publications, case studies, commissioning plans, sample reports, and other often-needed information. Two good Federal resources are the U.S. Army Corps of Engineers' *Commissionpedia - Electronic Sourcebook for Building Commissioning* and the General Services Administration's *Project Planning Tools*. The Califiornia Commissioning Collaborative also provides this information and more. All resources are listed in the references at the end of this chapter.

list of DDC inputs and outputs, their associated component, sensor or actuator type, and alarm limits. Sequences of operation should be updated for each HVAC and lighting system, with the rationale for any changes and deviations hi-lighted for the Operator's understanding. Updated system diagrams also assist the operators and maintainers by depicting the entire system in schematic format rather than in component bits and pieces.

O&M Staff Training

Training should actually be hands-on and take place throughout the recommissioning process.
The CxA provides additional training, or coordinates and arranges for training with appropriate vendors in conjunction with the recommissioning, if the improvements made to the equipment and systems warrant additional familiarization or emphasis or if a particular operation or practice is of a particular concern to the Owner. Since staff members who lack systems expertise are often the root cause of many problems that necessitate recommissioning services in the first place, the training should be given high priority.

PERIODIC REVIEW

Documentation Requirement:
◆ First-Year Report

The CxA performs follow-up site visits and interviews at periodic, prescribed intervals with Facility Management and O&M staff to review the system operation, identify any operating problems, and recommend further improvements.

Energy data is reviewed periodically, if not continuously, to assess the need for further recommissioning. At this time, increased building energy consumption and decreased performance efficiency is flagged. The CxA then works with the O&M staff to perform an evaluation, de-

References

1. *Retro-Commissioning's Greatest Hits*, Haasl, Tudi et al, Pacific Northwest National Laboratory, March 2000.

2. *A Practical Guide for Commissioning Existing Buildings*, Haasl, Tudi and Terry Sharp, April 1999.

3. *Energy Star Buildings Manual*.

4. *Commissionpedia*, U.S. Army Corps of Engineers; available online at www.cecer.army.mil/KD/HVAC/index.cfm?chn_id=1136.

5. *Project Planning Tools*, U.S. General Services Administration; available online at www.projectplanningtools.org.

6. *State of California Guide to Commissioning Existing Buildings*, California Commissioning Collaborative, February 2006; available online at www.CACX.org.

velop measures to restore the building energy and operational performance, and implement the measures.

Building staff play a key role in tracking the measures after they have been implemented to ensure that they work properly. This avoids the need for the CxA to make a return visit to verify that savings are persisting.

A first-year report is developed by the CxA. It documents measured energy savings, recommends any additional changes or building improvements, provides recommendations for ongoing O&M staff training, and establishes a schedule for future recommissioning, if a continuous commissioning program is not established. The more frequently the building undergoes changes in facility use and occupancy patterns, the more often the building should be recommissioned.

BEST PRACTICES

- Employ recommissioning to improve facility efficiency and address issues that may have arisen since the original commissioning.

- Keep recommissioning in mind when performing initial building commissioning: the more forms and documentation you can provide from the initial commissioning process, the better.

- Recommission the entire building every three to five years, or put in place an approach in which one-third to one-fifth of the building's systems are recommissioning each year, on a rotating basis. This makes recommissioning "business as ususal" rather than a special event.

- The more substantial the changes a facility endures, the more frequently it should be recommissioned.

- Integrate recommissioning into the facility's preventive maintenance program to improve the performance of both approaches.

- When recommissioning with a limited budget, focus first on HVAC, refrigeration, and electrical systems and their controls, which are often the sources of the biggest operational problems.

- Focus also on low- and no-cost recommedations, as these can have a surprisingly big impact on the facility's bottom line.

121

STUDY QUESTIONS

1. What building systems are most often involved in recommissioning, and why?

2. How long after initial building commissioning should one wait to recommission the facility? What factors go into determining how long to wait before recommissioning?

3. How does recommissioning differ from retrocommissioning? How are the two processes similar?

4. How can recommissioning be integrated with an existing preventive maintenance program?

5. What are typical items to look for during the design review phase of the recommissioning process?

6. How important is it to establish a performance baseline, and what are some typical methods for baseline development?

7. What role does TAB play in recommissioning?

Chapter 8
Continuous Commissioning Process

Continuous commissioning is a form of remote intelligence. The primary focus of continuous commissioning is ensuring the persistence of building systems optimization. It is an ongoing process for existing buildings, employed to resolve operating problems, improve building comfort and safety, optimize energy use, and improve system reliability.

As in recommissioning, continuous commissioning takes place only when the facility has been previously commissioned, since it needs a baseline for comparison. Its objectives are essentially the same – identifying and correcting building system problems and optimizing systems performance and reliability in existing buildings. The major difference between recommissioning and continuous commissioning is in the degree of persistence.

A typical continuous commissioning approach can be viewed as having four distinct functional processes as shown below:

■ The first of the functional steps is to monitor the building system(s) or subsystem(s) and **detect any abnormal conditions** – the fault detection phase.

■ If an abnormal condition is detected, then fault diagnosis is used to evaluate the fault and **diagnose the cause** of the abnormal condition.

■ Following diagnosis, during the fault evaluation phase, the **magnitude and impact** of the fault on factors such as energy use, system reliability, and plant operations is determined.

■ Finally a decision is made on **how to react** to the fault.

In many cases, detection of faults is relatively easier than diagnosing

123

A resting body tends to stay at rest and a body in motion tends to stay in motion unless acted upon by an outside force.

Law of Physics

> This chapter describes the process for implementing continuous commissioning in an existing facility. Chapter 5 covers the commissioning process for new construction and major renovation, Chapter 6 the retrocommissioning process, and Chapter 7 the recommissioning process.

124

While commissioning focuses on bringing the building operation to the design intent, continuous commissioning focuses on optimizing HVAC system operation and control for the existing building conditions.

them or determining their impacts. For those familiar with Reliability Centered Maintenance (RCM), the process of detection, diagnosis, risk analysis, and outcome is similar to RCM's decision logic tree.

This process is both "continuous" (uninterrupted) and "continual" (recurring regularly and frequently). Ensuring that building systems remain optimized continuously requires:

■ Benchmarking of energy, operations, and output data,
■ Continuously gathering new data,
■ Making comparisons between that new information and the benchmark data and against pre-established metrics and trends, and
■ Establishing new baselines.

Continuous commissioning also involves finding opportunities that will make the building run better and to its maximum efficiency without sacrificing occupant comfort requirements. Designers historically have oversized HVAC units by as much

as 50 percent to allow for contingencies and possible growth. The end result is oversized units running at inefficient part loads. Continuous commissioning seeks to mitigate those situations.

Continuous commissioning is closely related to (and often integrated into) a facility RCM program. By measuring output parameters (e.g., temperatures, pressures, volumes) and trending them over time or tracking them relative to alarm limits, impending problems can be investigated and averted.

The subtle difference is that instead of emphasizing a system or component's current condition and predicting an impending failure like RCM, continuous commissioning emphasizes optimal building and systems operation to meet current output requirements. Where RCM is heavily reliant on predictive technologies, continuous commissioning incorporates the permanent installation of metering equipment, software and building automation sensors. (RCM also incorporates this available output information as one of its many analytical tools.) On-going periodic (hourly) output and energy metering, monitoring, and analysis is conducted automatically to detect proactively when some event, deficiency, or impending failure is impacting a system's efficiency.

Continuous commissioning is reported to have produced typical savings of 20 percent with payback under three years (often one to two years) in more than 130 large build-

Continuous Commissioning Process

Project Development
Decide which equipment and systems will be
analyzed for potential improvements
Conduct commissioning audit and develop project
scope
Hire CxA or assign duties to O&M personnel
Develop detailed work plan and form project team

Implementation and Verification
Develop performance baselines
Measure current equipment and system performance and
develop recommended improvements
Track, trend and analyze energy and performance data for
anolmolies
Implement recommended improvements
Document energy and operational savings and comfort
improvements

Periodic Review
Continue to track, trend and analyze energy and
performance data for anolmolies
Follow-up site visits and interviews
Write follow-up project report at 1 year

Process begins again when
periodic review indicates
that the building,
equipment, and/or systems
are not operating at optimal
levels (anomalies are
found)

ings according studies by Texas A&M University.

PROJECT DEVELOPMENT PHASE

Documentation Requirement:
◆ Project Work Plan

The first step of continuous commissioning is performing an assessment of whether continuous commissioning makes sense. This involves conducting a needs assessment where a walkthrough is made by Building Management to identify and discuss the Owner's expectations for comfort performance, building energy performance, and known problems. It also involves a facility survey, review of equipment performance data and histories, and an assessment of the available automated system tools, which are essential for the continuity of the commissioning program, and their capabilities.

These automated systems are typically a facility's building automation system (BAS) or energy management and control system (EMCS), which provide local, remote-capable, system alarm, control, and archiving information. They allow Facility Management to monitor, manage, and control mechanical systems and lighting remotely. Consumption and control parameters are trended and compared. Alarms are indicated, usually communicated to a pager or cell phone, and a service technician responds as necessary. Some sites incorporate an enterprise energy management system (EEMS), which uses the Internet to connect all major energy-consuming devices in a facility or group of facilities, for additional analysis and problem solutions.

If it is found that continuous commissioning has merit, the process continues with the formation of the

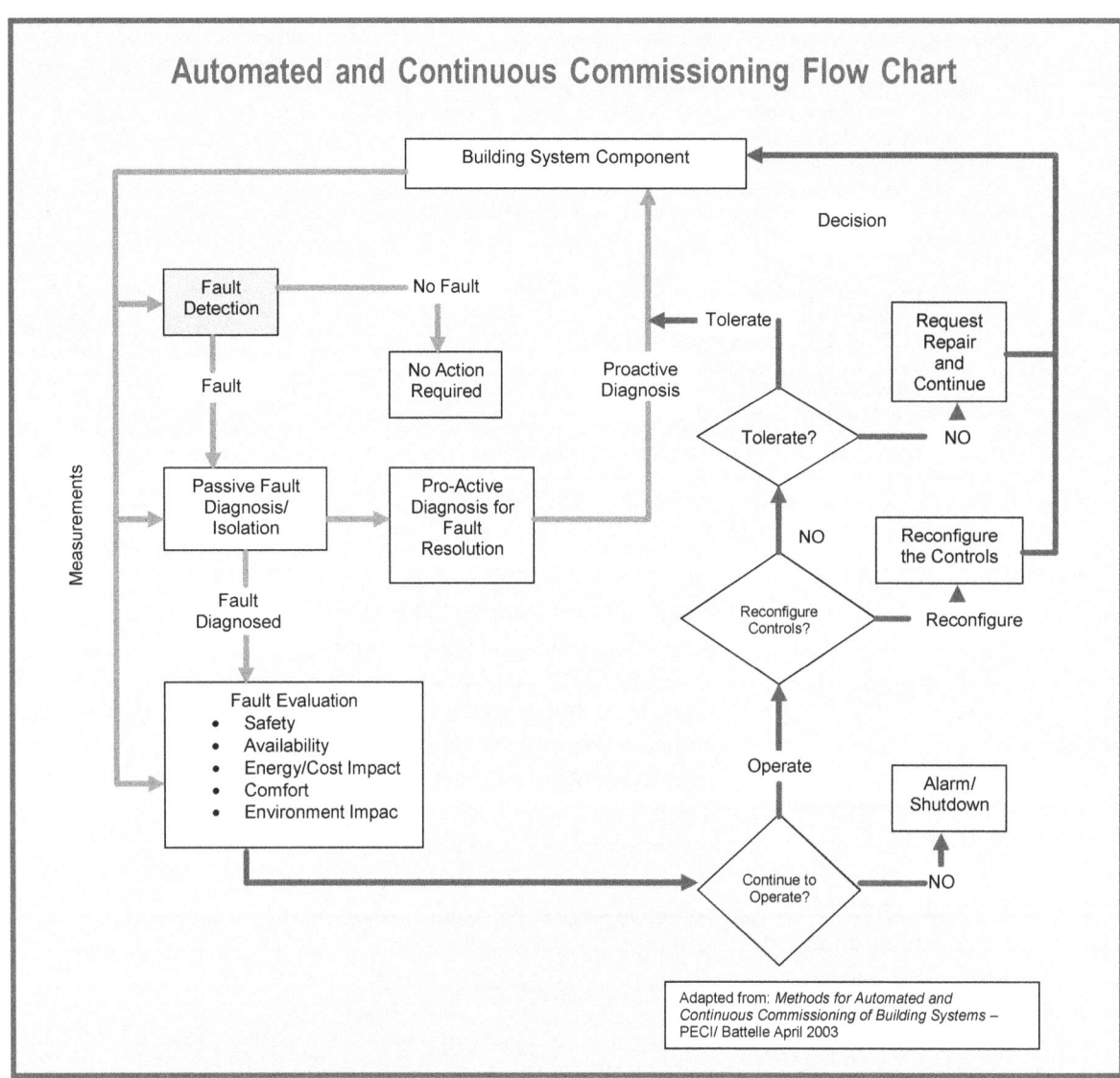

Automated and Continuous Commissioning Flow Chart

Adapted from: *Methods for Automated and Continuous Commissioning of Building Systems* – PECI/ Battelle April 2003

commissioning team and development of the work plan.

Commissioning Team

Continuous commissioning is usually implemented and conducted in-house with facility O&M staff, or the data can be transmitted electronically to an independent analytical laboratory where it is captured, stored, and analyzed on a subscription basis.

The continuous commissioning team consists of at least three individuals: the Facility Manager or Owner's Representative (Commissioning Authority); a systems engineer (Commissioning Engineer); and a systems technician (Commissioning Technician). Each has a distinct role in the continuous commissioning process:

Commissioning Authority / Facility Manager

- Prepares a team charter and ensures responsibilities and authorities are properly reflected in the appropriate position descriptions (contract document if the function is outsourced).
- Coordinates the activities of the commissioning team.
- Establishes team objectives relative to the Owner's requirements.
- Determines metrics to be achieved.
- Establishes and maintains schedules and milestones.
- Reviews results and approves remedial actions within his or her authority.

- Provides information / commissioning reports to Management, obtains "buy-in" and approval, and provides feedback and guidance back to the team.
- Follows up on the implementation of recommended changes.

Commissioning or Systems Engineer

- Develops automatic monitoring and field measurement plans.
- Establishes and maintains system performance baselines.
- Monitors and interprets data and trends.
- Diagnoses the causes of faults and their impacts.
- Conducts engineering analyses.
- Develops system improvement measures.
- Develops improved sequences of operation, control schedules, and set-points.
- Directs programming changes to the BAS/EMCS/EEM software.
- Estimates potential energy savings and costs to implement recommended measures.
- Develops implementation schedules.
- Provides guidance to the technicians implementing recommended changes.
- Documents findings and periodically updates the commissioning reports.

Commissioning or Systems Technician

- Provides input based on knowledge and experience, and building, system, and equipment histories to the engineer to be

127

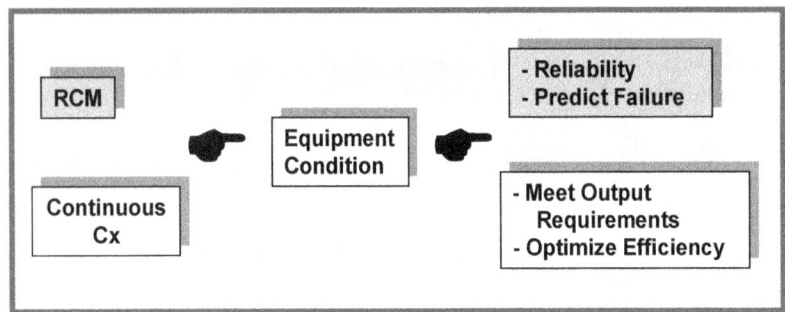

considered during the commissioning analyses.

- Conducts field measurements.
- Helps interpret collected data, as required.
- Implements mechanical, electrical, control, and systems software changes recommended by the Systems Engineer.

Work Plans

Appendices J and K of PECI/ORNL's *A Practical Guide for Commissioning Existing Buildings* provide good examples of detailed work plans. These plans document and maintain a focus on the systems being monitored and at what points, the devices performing the monitoring, their settings and locations to ensure continuity, minimum and maximum allowable parameters, and recommended actions in the case of non-compliance or alarm situations. They also include identification of involved personnel, general information narratives, special instructions, and step-by-step procedures, as appropriate.

Diagnostic monitoring, trending, and functional performance test plans are reproduced in Appendix A,

Sample Commissioning Forms at the end of this guidebook. Plans similar to these should be developed and used so that the effort stays focused, coordinated, and thoroughly documented.

IMPLEMENTATION PHASE

Documentation Requirements:
- List of Improvements Selected for Implementation
- Commissioning Report

The assessment may only identify suspected areas for improvement. There may be a need to obtain more complete and exact data on when and how systems are actually operating. Using the most recent baseline data from previous commissioning efforts, the commissioning team compares it to the actual collected data manually or automatically.

Step 1 - Fault Detection

Diagnostic monitoring allows the commissioning team to observe critical parameters such as inside and outside temperatures, humidity, flows, pressures, speeds, noise levels, light levels and intensities, and more under typical operating conditions. Three of the most common diagnostic monitoring methods are BMS/EMCS/EEM trend logging, standalone data logging, and manual functional performance testing.

BMS/EMCS/EEM Trend Logging

Facility performance data are gathered from the building's BAS or

EMCS using open protocols and software algorithms. Automatic trend logging can provide a wealth of system information using minimal human effort and with reasonable accuracy. It also can automatically alert when conditions approach abnormal.

However it is important that the sensors' calibration be maintained and that the user has confidence in the accuracy of the collected data. Another drawback of relying solely on the BMS/EMCS/EEM for diagnostics is that its sensors are permanently mounted and installed, and taking measurements at other locations is not possible.

Some sites have outsourced their BMS/EMCS/EEM functions to remote providers. Many of these have moved beyond alarm response and troubleshooting to trending and identifying system deficiencies and opportunities for improved efficiency. This information is then used to make better informed design decisions in the future.

Portable Data Logging

Portable data loggers are good for short term diagnostic and monitoring activities. These are small, battery-powered devices that are easily installed, often with magnets, and removed without disruption to the building occupant. While some indicate current real-time conditions, others are highly sophisticated and, when downloaded on a laptop or other device, will trend, graph, and analyze the collected data. Manual testing and spot observations do not provide this level of accuracy.

> ## Case Study: Continuous Commissioning in Action
>
> A medical center in Texas was commissioned in 1992, recommissioned in 1993, and produced savings of $145,000 and $62,500 respectively. Continuous commissioning techniques were applied in 1994 and obtained an additional $195,000 per year savings on the same systems that had been previously commissioned to design specifications. (Source: Culp, Charles H. et al, *Continuous Commissioning in Energy Conservation Programs*, http://energysystems.tamu.edu.)

129

Functional Performance Testing

Manual functional performance testing may be conducted to verify a system's correct operation and to pinpoint problems. It involves putting each system or piece of equipment through a series of tests that check the operation under various modes and conditions. The condition can be simulated or a mode of operation can be forced manually. Data are gathered by taking spot measurements using hand-held instruments such as multi-meters, ammeters, infrared cameras, and light meters. The data is then compared against the baseline, other diagnostic results, and/or design or desired requirements.

It is beneficial to conduct all three diagnostic methods simultaneously, if possible. For example, if the functional performance testing is conducted while data loggers are still in place, the results will be permanently recorded and can be graphed and analyzed. At the same time, the EMCS can be used to view the various responses as they occur.

Trend logs are vital tools to aid diagnostics. They frequently reveal unexpected results. Some recorded examples include:

- Boilers firing for short spurts during the summer cooling season for no apparent reason.
- Energy spikes occurring at about 2:00 a.m. each day over a month's time with no known cause.
- Control dampers and valves completely recycling every five or ten minutes.

Step 2 - Diagnosis

Semi-automated diagnostic tools are available that rely on data that is collected by the BAS and portable data loggers and then use internal algorithms to detect and diagnose problems in the equipment. They reduce the significant time burden of gathering, downloading, and converting data into a suitable format for analysis and reduce the analysis time as well. They also reduce the required skill level of the person conducting the testing, since most of the technical expertise is required up front in setting up and then maintaining the software. Examples of semi-automated diagnostic tools are:

Whole Building Diagnostician (WBD) developed by Pacific Northwest National Laboratory - The Whole Building Diagnostician has two modules: *Overall Energy Use* and *Economizer Optimization*. The Overall Energy Use Module calculates the total building energy use as a function of outside air temperature and other key parameters. Over time, when the total energy use exceeds a pre-established alert level, it alarms the user. The Economizer Module periodically measures the condition parameters of air flows, outside conditions, and status data to determine the operating state of AHUs. Both modules detect faulty or misplaced sensors and estimate energy and cost impacts of all problems found. The economizer module further identifies probable causes of the problem and recommends remedial measures.

Performance and Continuous Recommissioning Analysis Tool (PACRAT) developed by Facility Dynamics Engineering - The PACRAT provides trend data recorded and stored by a BMS/EMCS, data log-

---

Common Problems Detected by Diagnostic Tools

- ◆ Uncalibrated and failed sensors
- ◆ Simultaneous heating and cooling
- ◆ Leaking valves
- ◆ Unoccupied periods
- ◆ Inadequate ventilation rate
- ◆ Economizer opportunities
- ◆ Improper terminal outputs
- ◆ Setpoint deviations
- ◆ Incorrect of ineffective sequencing
- ◆ Poor efficiency
- ◆ Excessive cycling
- ◆ Excess energy use relative to the baseline

130

gers, or metering system. Comparing collected data to a baseline, it diagnoses system problems and poor performance, and identifies energy wastes. It integrates weather data as necessary. It includes "expert" advice on the possible cause of the anomaly and how to correct it. It identifies energy waste, why it is occurring, and suggests how to remedy it.

ACRx Palm Pilot tools developed by Field Diagnostic Services - The ACRx Palm Pilot tools allow for the field testing of HVAC equipment and controls. Using thermodynamic principles, these tools detect and diagnose HVAC problems, perform short term monitoring of HVAC parameters, and monitor long term controls performance.

AEC ENFORMA - This software programs data loggers to gather and process performance data. It assists in setting up optimum data points and provide the user with graphs of the collected data. Trend analysis remains primarily a manual process.

Manufacturers also provide diagnostic capabilities in many of their equipment controllers. For example, chiller controllers have long had diagnostic capabilities through a human interface and digital display that show operating and diagnostic codes, compressor status, setpoints, specified temperatures, specified pressures, and enable/disable features and options. Faults and operating conditions are identified and percent load and percent kW are displayed that indicate the loading condition and efficiency of the chiller.

A properly implemented continuous commissioning program will:

◆ Reduce energy usage and costs through monitoring and management

◆ Improve comfort, health, and safety of building occupants by maximizing the air quality of their working spaces

◆ Improve the condition and reliability of the building operating systems

◆ Facilitate compliance with the EPAct and current energy Executive Orders.

131

Baselines

The CxA refers to the original and any subsequent commissioning baseline data. Then, using data from the BAS/EMCS and other installed meters and data loggers, compares the results.

The baseline performance criteria is re-evaluated at least annually, since settings tend to drift away over time. Changed building uses, missions, occupant changes, configurations, and even weather patterns all will have some impact. For example, a previously unoccupied space may become occupied or the building may have had a mission change from an administrative center to a conference area. It is possible that due to these changes, the baseline may no longer even be achievable, the systems operation and efficiency no longer optimal, and higher energy costs may likely be the result.

In addition to the system performance and output data, it is also necessary to look at energy information. Comparative energy data may be accumulated using any one or a combination of the following:

- Short-term measured data from data loggers, BAS, or EMCS.
- Metered long-term hourly or 15-minute whole building energy data.
- Historic electricity, gas, and steam utility bills.

The Facility Manager reviews the energy data at least quarterly. If the building energy consumption has increased relative to established metrics and baselines, the systems engineer investigates to determine its possible causes, with input as needed from the operations staff.

The building annual Energy Use Index (EUI) is the common benchmark used to make building energy use comparisons. It is expressed in terms of energy use per square foot of floor area (Btu/ft^2 or kWh/ft^2). Increased energy use over time in the same building area may signal the need for an energy audit and remedial action. In addition, because it is normalized to the floor area, comparisons can also be made between similar buildings or buildings with similar functions, but with different floor areas.

Step 3 - Fault Diagnosis

Monitoring BMS/EMCS/EEMS data produces real-time information and specialized reports that permit the facilities staff to recommission buildings continuously. For example, typically the building operations staff would use the building's control sys-

tem to measure space temperatures and reduce duct static pressure gradually until the lowest static pressure that maintains occupant comfort is found. That setting would remain until it is again adjusted manually.

Continuous commissioning performs that same procedure, but with necessary adjustments, over and over again several times a day. Instead of finding a single optimum setpoint, the BMS/EMS/EEMS is used to find the most appropriate setting under changing conditions – morning versus afternoon, summer versus winter, and so forth. The end result is that energy usage is held in check without sacrificing comfort and health.

As anomalies, inefficiencies, and deficiencies are detected, the Commissioning Engineer develops a detailed measurement cut-sheet for each major system, listing all parameters to be more carefully measured and all parts to be more accurately checked.

Continuous commissioning does not overlook its most important resource – people. Occupant feedback, coupled with the results from data collection devices, may be able to shed some light on possible causes of building anomalies and deficiencies. It also identifies comfort problems in specific areas attributed to temperature, humidity, noise, odors, air flow, and lack of outside air. Information from the building maintenance staff reveals operational problems, recurring problems, and the need to frequently replace parts,

Continuous monitoring allows the Owner to observe energy consumption patterns and see how they change under varying conditions, thereby helping the Owner make informed utility purchasing decisions, such as peak shaving and interruptible service.

which is indicative of bigger problems.

In some cases, more detailed root cause failure analysis should be performed. Plant equipment may fail repeatedly or the work environment may always be uncomfortable. These failures eventually become accepted as a normal condition. Recurring problems, such as short bearing life, loud noises in the ductwork, and mold are symptoms of more severe problems. However, maintenance personnel often only fix the symptomatic problems and continue with the frequent repairs or adjustments necessary to ameliorate the problem for the moment. Repeated failures and uncomfortable conditions result in high costs for parts and labor and decreased occupant productivity. Further, unreliable equipment and poor indoor work environments may pose a continuing personnel health and safety hazard.

While continuous commissioning can identify equipment faults at such an early stage that they never lead to actual equipment failure, it often does not include discovering the underlying reason for the faults. For example, an air handling unit may not be delivering the correct quantity of conditioned air to a space. Commissioning may likely recognize that a belt is badly worn or broken and needs replacement. But if nobody recognizes that the sheave is bent and is causing premature belt wear or that the technician used a screwdriver to pry the belt over the sheave and bending it because he or she was not properly trained on replacing belts on that specific equipment, then the failures will recur at continued downtime, cost and discomfort.

Root Cause Failure Analysis (RCFA), a fundamental tool of RCM, can be used to proactively seek the fundamental causes that lead to equipment failure, excessive energy use, and poor working conditions. Its goals are to:

- Find the cause of the problem quickly, efficiently, and economically.
- Correct the *cause* of the problem, not just the effect.
- Provide information that can help prevent the problem from recurring.
- Instill a mentality of "fix forever."

Detailed instruction on how to conduct a RCFA is beyond the scope here. However, ample guidance is available through internet search engines on the topic of Root Cause Failure Analysis and/or Reliability Centered Maintenance.

Step 4 - Fault Reaction

Using these measurements and other input, the Commissioning Engineer and Commissioning Authority conduct an engineering analysis and

133

risk assessment, respectively, to determine their impacts on energy consumption, system reliability, mission, operations output, budget, and on other concerns to the Owner. A decision is made to either take immediate remedial measures (such as equipment shut-down or emergency repairs), to schedule remedial measures or improvements, or to accept the risk and tolerate the existing conditions. Remedial measures may be as simple as adjusting control schedules and setpoints or as drastic as making capital improvements to improve energy efficiency.

The approach selected should be based on an evaluation of the following factors:

- **Consequences of Failure** – Consider the impact of a failure (actual or impending) on safety, security, environment, health, mission, productivity, cost, and morale/political impact. If there is no impact, then there is no justification for monitoring the conditions at all.

- **Probability of Failure** – Consider the maintenance history and the reputation of the systems being monitored.

- **Redundancy** – Evaluate whether or not there is a single point of failure that can shut down operations.

- **Time Path for Failure** – Based on the operating environment, load, tolerances, and location, consider whether the projected time period between

the start of a system degradation and its functional failure is unacceptable.

- **Predictability** – The measurement of degradation is obvious to the system operator or user.

- **Cost** – Compare the relative costs of monitoring against the anticipated benefits.

- **Process Time** – Determine if the time between data collection and analysis is critical to operations.

- **Accessibility to Monitoring Locations** – For example, consider critical air handling units where the starting and stopping of fans would be required to safely and accurately take measurements. This situation is a good candidate for remote monitoring.

Monitoring and Verification

The impact of the minor adjustments made during the functional performance testing and of any implemented improvements are monitored immediately following their implementation to verify their impact on occupant comfort, on operational and energy performance, and on control sequences and schedules.

Room-by-room measurements are made using hand-held meters or portable data loggers. Energy model data is collected using the short-term methods established earlier. Utility bills and meter data for electricity, gas, steam, and Btus are reviewed for long-term monitoring. The outcome is updated details on energy consumption, use patterns, trends, and new baseline data.

If the monitored data shows no abnormalities or disturbing trends, it is good practice for the building manager to check with key occupants and maintenance personnel semiannually to ensure that there are no comfort or operational problems being experienced that are not otherwise detected by the BMS/EMCS/EEM.

Implementation of Improvements

During each recommissioning cycle, a report is prepared by the commissioning engineer and approved by the commissioning authority that summarizes the results of the energy performance baseline measurements and the system measurements. The purpose of the report is to provide data interpretation and action plans – not just performance charts and graphs.

An executive summary includes a summary of participants and their roles, building description, project objectives, an overview of the commissioning scope, and a general description of testing methods. The report includes lists of existing control sequences and set points for all major equipment, disabled control sequences, and malfunctioning equipment and control devices.

The report addresses performance relative to pre-established metrics. It includes a list of recommended improvements and operation and control procedures and their estimated impact on performance and

135

on implementation costs and savings. This report is presented to the Owner or Owner's representatives to obtain approval to implement recommendations beyond the authority of the CxA and/or Facility Manager and to obtain their "buy-in" and ongoing program support.

Upon approval by the Owner, the commissioning team works with the building O&M staff and any vendors, as necessary, to implement the selected cost-effective improvements. **The highest priority is to solve existing problems.** Measures that optimize (not correct) equipment and systems and improve operation and control schedules follow next. Life-cycle cost analyses are performed, if possible, to identify where the biggest benefits relative to cost lie. While making improvements, the commissioning team ensures that system documentation is updated to reflect changes, such as to control sequences.

An improved and a more consistently managed indoor air environment leads to a more comfortable and healthier workplace. The result will be increased productivity. Also, since monitoring system operations and output parameters contributes to the equipment condition data under an RCM program, system reliability and maintainability, and its associated costs, will also improve.

One drawback is that improving indoor air quality and comfort may actually increase building energy consumption. For example, increasing the minimum building outside air will increase the air-conditioning load in the summer and the heating load in the winter. This trade-off must be reconciled during the commissioning analysis.

O&M Staff Training

Since continuous commissioning is likely to be an in-house function, it is necessary for the responsible personnel to be trained in and to have a good understanding of the BAS/EMCS software and monitoring equipment capabilities, data analysis, data collection techniques, and standard operating procedures (SOP) to be followed when disturbing data and trends are detected.

BEST PRACTICES

- It may be easier and in the long run more cost effective to hire an independent, third-party Commissioning Authority to set the commissioning project in motion and to train and mentor the in-house staff as needed until the comfort and technical levels required of the staff are achieved.

- Ongoing monitoring and verification of energy and operational performance are essential to maintain persistence of improvements and to ensure that equipment and systems are operating at optimal efficiency.

- A well-trained O&M staff with adequate resources is crucial to the success of any commissioning program.

References

1. *Retrocommissioning Handbook for Facility Managers*, Portland Energy Conservation, Inc., March 2001.

2. Liu, Mingsheng, Ph.D., P.E., et. al., *Continuous Commissioning Guidebook: Maximizing Building Energy Efficiency and Comfort*, October 2002.

3. Haasl, Tudi and Terry Sharp, *A Practical Guide for Commissioning Existing Buildings*, April 1999.

4. Culp, Charles H., et al, *Continuous Commissioning in Energy Conservation Programs*, http://energysystems.tamu.edu).

5. Mahling, Dirk K. and Lehman, Keith, *Enterprise Energy Management*, HPAC Engineering, December 2005.

6. *Methods for Automated and Continuous Commissioning of Building Systems*, Portland Energy Conservation, Inc. and Battelle Northwest Division, April 2003.

- Equipment, sensors, and data collectors must be accurately calibrated to ensure the validity of the collected data.

- Problems, deficiencies, and complaints should be carefully recorded by the facility Management and O&M staff. Often, problems can be grouped into categories (design, operation, maintenance, installation, comfort and safety) that can be analyzed for more significant trends.

- The documentation required by each commissioning process can be daunting, but will prove an invaluable resource for building management, O&M, and future commissioning efforts.

STUDY QUESTIONS

1. What are the documentation requirements for each phase of the continuous commissioning process? Why are these documentation requirements important to the overall success of the continous commissioning process?

2. How does continuous commissioning differ from new construction/renovation commissioning? How does it differ from a typical preventive maintenance program?

3. Who are the continous commissioning team members, and how do their roles differ from a new construction/renovation commissioning team?

4. What are the four steps involved in the implementation phase of continuous commissioning?

5. What are the methods of investigation used during the continuous commissioning implemention phase?

6. How and why should energy and performance baselines be established?

7. What are the factors involved in developing an approach to correcting detected anomalies?

8. How would you prioritize and implement improvements recommended through the continuous commissioning process?

138

This page left blank intentionally

Chapter 9
Sustainable Commissioning

A successful commissioning program takes thorough planning and expert execution, but several additional considerations are important to the success of commissioning, including developing a good measurement and verification program and ongoing operations and maintenance staff training.

In addition, the Federal government has demonstrated a commitment to sustainable design, and commissioning has a central role to play in the building and maintenence of "green" buildings as well.

COMMISSIONING FOR
LEED CERTIFICATION

This document was developed by the U.S. Department of Energy (DOE) Federal Energy Management Program (FEMP) to help Federal facilities managers, their staffs, and their consultants understand the basics of commissioning and to apply its practice to opportunities within their own organizations. The Federal facilities managers, staffs and consultants toward whom these materials are directed manage several hundred thousand facilities worldwide in about three billion square feet of floor space. They

purchase billions of dollars of materials for operations, maintenance, repair, and renovation.

The collective impact that commissioning could have on building operating and maintenance costs, energy usage, and environmental impacts is tremendous, and this potential is not going unrecognized. In fact, commissioning – as has been discussed throughout – is now a mandatory requirement for the acceptance of many sustainable or "green" buildings.

And this leads us to LEED – Leadership in Energy and Environmental Design.

The LEED Green Building Rating System™ is a voluntary, onsensus-

Don't blow it - good planets are hard to find!

Environmental Corollary

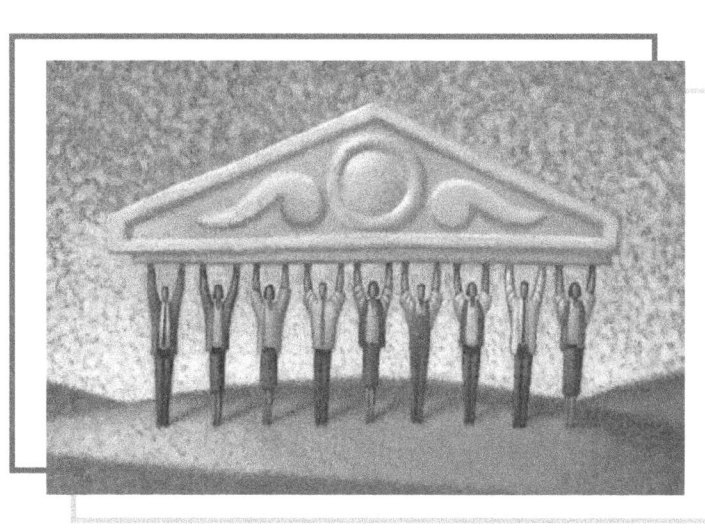

In this Chapter

◆ Commissioning for LEED Certification

◆ Measurement and Verification of the Commissioning Program

◆ Post-Commissioning Training Program

◆ Best Practices

140

LEED certification aims to improve occupant well-being, environmental performance, and economic returns of buildings using established and innovative practices, standards, and technologies.

based national standard for developing high-performance, sustainable buildings. Administered by the U.S. Green Building Council, the LEED program represents all segments of the building industry, and provides standards for new commercial construction and major renovation projects, existing building operations, commercial interiors projects, core and shell projects, homes, and neighborhood development. It has further developed guidance for specific applications in retail, multiple buildings and campuses, schools, healthcare, laboratories, and lodging.

LEED provides a complete framework for assessing building performance and meeting sustainability goals. The certification program emphasizes state-of-the-art strategies for sustainable site development, water savings, energy efficiency, materials selection, and indoor environmental quality.

Understanding the LEED certification ratings and how to achieve facility certification with design, construction, and operational credits based on the system has become

imperative as government agencies look for ways to become more environmentally friendly, conserve energy, and decrease the operating costs of their real estate. Indeed, current Federal Laws, Executive Orders, and Executive Memoranda direct Federal Government facility managers to reduce the energy and environmental impacts of the buildings they manage.

Commissioning all building energy systems is a prerequisite for every LEED project. Any building can benefit from commissioning. However, it is even more important in energy efficient buildings as advanced control strategies become increasingly complex. Beyond HVAC, commissioning has become invaluable to assessing the integrity of the building envelope to ensure comfort, weighing in on manufacturer claims of building materials appropriateness, and in confirming any equipment condition relative to possible latent manufacturing, transportation, and installation defects.

The USGBC firmly believes that third-party commissioning is invaluable to the building Owner – so much, in fact, that in addition to it being a requirement for certification, additional points can be attained by conducting optional commissioning activities. This commissioning, as discussed earlier, starts during the planning and design stages, through construction, and well into the post-construction and warranty phases. LEED requires the commissioning of the facility's static and dynamic elements, particularly any that affect energy efficiency and the indoor environmental quality.

Commissioning Components

Specific prerequisite and optional commissioning requirements can be found at the USGBC LEED website (www.usgbc.org) and will not be repeated here to maintain currency with the constantly evolving applications. However, in general, LEED requires commissioning, including thorough documentation, of building integrated HVAC systems and controls, ductwork and pipe insulation, renewable and alternative energy technologies, lighting controls and daylighting systems, waste heat recovery systems, and other advanced technologies. It also includes testing, adjusting, and balancing (TAB) verification. Certain site features, such as alternative fueling stations and exterior lighting, are also required to be commissioned for LEED points. Water systems such as irrigation systems, plumbing fixtures, and plumbing infrastructure are also commissioned under LEED.

To demonstrate the successful completion of the commissioning requirements, the applicant must provide a copy of the project's Commissioning Plan that highlights each of the fundamental best practices commissioning procedures used, and a signed letter of certification from the CxA confirming that the Commissioning Plan has been successfully executed and the de-

LEED Certification Benefits

Often cited reasons for building owners and facility managers to design, construct, and manage their facilities to meet LEED criteria:

◆ Easier operation and maintenance that meet the user's constantly changing needs

◆ Bragging rights for high performance buildings that are "Showcase Facilities"

◆ Superior indoor air quality

◆ Comfortable indoor environment with emphasis on thermal, visual, and noise conditions

◆ Improved system reliability

◆ Efficient use and selection of materials and supplies to maximize recycling and minimize disposal

◆ Maximized use of natural sunlight

◆ Conservation of water and minimized water waste

◆ Thorough documentation and performance baselines

◆ Maximized energy efficiency and use of renewable energy

141

sign intent of the building has been achieved.

The additional commissioning credits focus on reviews of the building design and construction documents to identify potential problems and opportunities for improvement, on establishing a program for future-year recommissioning, and in continued energy measurement and verification after turnover to help the Owner manage and recoup savings.

Documentation required from these activities include an excerpt from the commissioning plan describing these activities, a copy of the design reviews, and a signed letter from

Case Study: Potential Cost Savings of LEED Certification

The Costs and Financial Benefits of Green Buildings, A Report to California's Sustainable Building Task Force (Greg Kats, Oct 2003) identified cost savings resulting from pursuing LEED certification on 33 buildings. The report identified a financial benefit of $50 to $70 per square foot over a 20 year life of these buildings. An estimated $8.47 per square foot is attributed to O&M savings resulting from commissioning. Factoring in the implementation costs of the sustainable design concepts, there was a 10-to-1 payback. (Source: Berning, Michael, *Commissioning for LEED Projects*, Engineered Systems, January 2006.)

142

the CxA stating that all of these tasks were completed successfully. For the continued energy measurement and verification, the Owner must further comply with the DOE M&V procedures, provide an M&V plan, identify a schedule of instrumentation and controls input and output data to be collected, and include cutsheets of the sensors and data collection devices that will be used.

MEASUREMENT AND VERIFICATION OF THE COMMISSIONING PROGRAM

Adopting measurement and verification (M&V) approaches will help indicate the effectiveness of the commissioning program, both as it unfolds and at project completion.

The easiest way to quantify equipment and system performance before and after retrocommissioning is

by measuring energy performance. A problem that is identified and corrected may result in reduced trouble call time for your O&M staff, but this savings is hard to assign a dollar value to, and may not be noticed until well after the retrocommissioning project is complete.

The establishment of an annual energy use index (EUI) provides a baseline to estimate energy savings when proposing a retrocommissioning project, and to calculate actual savings after project implementation. The baseline should be sophisticated enough to differentiate between only those energy reductions that result from the retrocommissioning project and not reductions that occur from changes in building use, weather, etc., and should be flexible enough to accommodate changes that occur after the project is underway.

A thorough energy analysis of your facility will:

- Evaluate and describe all energy end uses.
- Identify energy consumption by system and fuel type.
- Summarize the operation schedules of systems.
- Describe the efficiency of all systems in your facility.
- Identify operation, scheduling and maintenance efficiency opportunities.
- Describe opportunities for off-the-shelf efficiency technologies.
- Describe engineered energy efficiency solutions.

- Describe in detail the cost and energy cost savings of any capital investments recommended.
- Identify opportunities that will require further design.

The three basic methods for establishing an EUI baseline are:

1. *Energy Calculations*, which incorporate information about and energy consumption history of energy-using building systems and equipment.

2. *Regression Analysis*, a statistical technique that uses historical data derived from meters to isolate one or more variables that affect energy use (resulting, for instance, in an equation that relates energy use to weather or building use variables). When historical, metered data are available, regression analysis defines energy use relative to the entire building and allows greater flexibility in making recommendations related to energy efficiency.

3. *Simulation*, a sophisticated set of engineering calculations that attempts to forecast energy use on the basis of a building's size and shape, equipment, levels of insulation, types of windows and doors, etc.

There are many software packages available to help your facility develop an EUI for benchmarking improvements, including DOE2, Trane TRACE, Carrier HAP, BLAST, and Energy Plus.

There are also free resources available online, such as the benchmarking spreadsheets for office buildings provided by the Oak Ridge National Laboratory. The benchmarking spreadsheets provided allow you to identify where your specific office building ranks relative to others. They calculate the EUI of your building, provide the typical (median) EUI for office buildings with the same characteristics as yours, and identify where your building's performance ranks compared to others (percentile of EUI).

The benchmarking spreadsheets go beyond the customary normalization by floor area and account for performance differences due to variations in worker density, the number of personal computers, operating hours, occupancy type, and heating fuel types. Beyond floor area, these characteristics were found to be the most common and most important drivers of electric and non-electric energy use in U.S. office buildings. Location effects are accounted

Commissioning is now a mandatory requirement for the acceptance of many sustainable or "green" buildings, and is a prerequisite for LEED certification.

144

for by specifying the census division corresponding to your building location.

In addition, the EPA's ENERGY STAR Label for Building also offers an online benchmarking tool called Portfolio Manager. Portfolio Manager can generate a Statement of Energy Performance for any building in your portfolio. This document communicates information about a building's energy performance in a format that is both understandable and easy-to-use in business transactions. The Statement of Energy Performance can help you formalize performance expectations to support leasing, building sales, appraisals, insurance, staff management, and commissioning, energy, and O&M service contracts.

Also, DOE's Federal Energy Management Program offers the Building Life-Cycle Costs (BLCC) software programs can help you calculate life-cycle costs, net savings, savings-to-investment ratio, internal rate of return, and payback period for Federal energy and water conservation projects funded by agencies or alternatively financed. The BLCC programs also estimate emissions and emission reductions. An energy escalation rate calculator (EERC) computes an average escalation rate for ESPC contracts when payments are based on energy cost savings.

To measure and verify other building performance data (apart from energy), consider using a performance metrics and benchmarking software program like Metracker, a prototype computer tool designed to demonstrate the specification, tracking, and visualization of building performance objectives and their associated metrics across the complete life cycle of a building (developed by the Lawrence Berkeley National Laboratory). Performance objective metrics established during pre-design planning can be used to guide and evaluate design decisions and can be updated to reflect the intended performance of the final design. These design intent metrics can then be used as benchmarks during commissioning and updated again to act as benchmarks for O&M diagnostics. The history of building performance documented can ultimately be used to better plan for, design, and operate future buildings. Such a methodology may also prove to be useful in documenting and tracking compliance with emerging commercial building rating systems.

POST-COMMISSIONING TRAINING PROGRAM

An essential element to the sustainability of the results of any commissioning effort is the proper training of the operations and maintenance (O&M) staff. The O&M staff must be thoroughly trained on how new or renovated equipment and systems are designed to work (design intent), how to properly operate and maintain them, and how to maintain operational and energy efficiency results through continuous commissioning approaches.

There are several common problems that hinder this vital process, however. Although O&M staff training is cited as a requirement for most commissioning projects, the actual training is often an afterthought as the project nears completion. Too often training is poorly coordinated, loosely-structured, informal, lacking in well-defined objectives, and too tightly scheduled. Training is often handed over to equipment manufacturers, who can certainly provide adequate information on their own products but have no idea about the overall building design intent and how their equipment should operate in concert with the entire building.

To be truly effective, the Owner and CxA must work together to develop and implement an O&M training program that is ongoing, thorough, and adaptable. Training must be thought of as a dynamic, open-ended component of the commissioning process, not a one-time event at the conclusion of the commissioning process.

Types of Training Programs to Implement

A comprehensive training program will involve different types of training at different stages in the commissioning process. The best program will provide an integration of ongoing O&M training in commissioning and preventive maintenance/RCM approaches. Typical training program components will include:

1. Initial training on new or renovated equipment and systems upon completion of the commissioning process. This training is optimally provided by the CxA or each responsible installing contractor to cover equipment and systems in all relevant areas: mechanical, electrical, HVAC, controls, etc.

   Oftentimes training is left to the equipment manufacturers. Although this approach can work, it is essential that the CxA reviews the training materials and provides supplementary training in systems and equipment integration with the rest of the building.

   It is highly recommended that all training during the project completion process be professionally videotaped; this will provide both a good guide to subsequent new hires as well as an instant "refresher" course for the future.

2. Initial systems training subsequent to project completion for new hires.

145

3. Periodic "refresher" training for the entire O&M staff. Refresher training, scheduled for perhaps once a year, is also a good opportunity for the Facility Manager to review any changes to equipment and systems and implement any new training required to address these changes.

4. Proficiency training to improve overall O&M staff expertise.

5. Cross-training to provide staffing flexibility and encourage the entire O&M staff to think of the building's systems and equipment holistically, rather than focuses solely on their area of expertise.

6. Certification and re-certification training to meet external regulatory requirements.

An integrated and ongoing systems approach to O&M staff training will provide existing staff with the knowledge base they need to properly operate and maintain complex and interdependent building systems, as well as provide a means to pass that knowledge base on to succeeding O&M staff.

BEST PRACTICES

- Commissioning should be seen as an ongoing process, not a distinct event, that will reap the most benefits if carefully planned and implemented for the life of the building.

- Just as commissioning is a process and not an event, training of the building's operations and maintenance staff must also been viewed as an ongoing process.

- Videotape training sessions to maintain a record for future generations of O&M staff, and review training requirements and materials at least once per year to ensure their relevancy and appropriateness.

- Some landlords, who recognize the benefits of meeting the stringent LEED criteria but do not want to to pay associated LEED expenses and are willing to forego the LEED certification recognition, still specify compliance with the LEED criteria in their Design Intent and Basis of Design. Doing so (and building to it) ensures that their building will still be commissioned and will have the quality and reliability benefits that go with it, that it will be energy efficient, and that

References

1. *Greening Federal Facilities – An Energy, Environmental, and Economic Resource Guide for Federal Facility Managers and Designers*, National Renewable Energy Laboratory, 2nd Edition, May 2001.

2. "Elements of an Effective Post-Commissioning Training Program," Sebesta Blomberg & Associates, 2002.

3. Berning, Michael, *Commissioning for LEED Projects*, Engineered Systems, January 2006.

4. USGBC LEED Application Guides available at http://www.usgbc.org/DisplayPage.aspx?CMSPageID=276&.

it will be environment-friendly without some of the expenses and management requirements associated with the actual certification.

■ Commissioning should always be included as a mandatory project line item and should not be allowed to be value-engineered out. The cost of commissioning is very small relative to its benefits and the overall project cost. Allowing it to be value-engineered out is similar to foregoing all the quality control checks that a new car goes through. If you needed to stay within the budget on a new car purchase, isn't it wiser to forego the leather seating or chrome wheels rather than delete the testing of its integrated systems to make sure that they operate as they should?

STUDY QUESTIONS

1. What is LEED certification and why is it so important to Federal facility owners and managers?

2. In general, what are the commissioning requirements for LEED certification?

3. What are some of the methods typically used to develop an annual energy use index (EUI)? What is the benefit of developing an accurate EUI for your facility?

4. What are some of the resources available to help when trying to measure and verify building performance?

5. What are some of the mistakes typical of an incomplete O&M training approach?

6. How often should O&M staff receive training on commissioned systems?

148

This page left blank intentionally

www.ingramcontent.com/pod-product-compliance
Lightning Source LLC
Chambersburg PA
CBHW080640180526
45168CB00008B/3241